The Compleat Strategyst

Being a Primer on the Theory
of Games of Strategy

J. D. WILLIAMS

With Illustrations by Charles Satterfield

DOVER PUBLICATIONS, INC.
NEW YORK

This Dover edition, first published in 1986, is an unabridged and
unaltered republication of the 1966 revised edition of the work first
published by the McGraw-Hill Book Company, New York, in 1954. It is
reprinted by special arrangement with The RAND Corporation, 1700
Main St., Santa Monica, California 90406-2138.

Manufactured in the United States of America
Dover Publications, Inc., 31 East 2nd Street, Mineola, N.Y. 11501

Library of Congress Cataloging-in-Publication Data

Williams, J. D. (John Davis)
   The compleat strategyst.

   Reprint. Originally published: New York : McGraw-Hill, 1966.
   Includes index.
   1. Games of strategy (Mathematics)   I. Title.
QA270.W5   1986   519.3   86-1067
ISBN 0-486-25101-2

To Evelyn
who even read it

. . . nevertheless, I believe that in considering these
things more closely the reader will soon see that it is
not a question only of simple games but that the foundation
is being laid for interesting and deep speculations.

—*Huygens,* 1657

Preface to the Revised Edition

I have been asked from time to time over several years by Professor A. W. Tucker whether I intended to do something further with the *STRATEGYST*. Inasmuch as each query was voiced as if he had never asked before, I was able for some time to use the same economical answer, "No." However, it came to me recently that he may have been trying, gently, to tell me something, so I changed my answer to, "Why?" It turned out that the *STRATEGYST* was, in an important particular, obsolete.

The important particular is the existence of what I shall call simply the pivot method, a general method for solving matrix games, and one which contrasts strongly with the collection of piecemeal methods used in the *STRATEGYST*. The method is a natural—almost an expected—consequence of the intensive development and convergence of the fields of game theory and linear programming, both pursued so assiduously since the mid-1940's, especially by groups at Professor Tucker's institution and at mine, Princeton University and The RAND Corporation. It is an outgrowth of the Simplex Method of my former colleague, Professor G. B. Dantzig, now of the University of California at Berkeley. Many distinguished mathematicians have contributed to the development, but the present format is essentially that of Professor Tucker, who has developed a combinatorial linear algebra of great generality and power. It deals in a unified way with many phenomena, including matrix games and linear programs. However, for the benefit of lay readers in the field of matrix games, who need simple arithmetic procedures, I have taken a step backward from theoretic unity—and elegance—by requiring that matrix elements be non-negative. The connections between the methods used early in the book with those of Chapter 6 are established by Tucker in "Solving a Matrix Game by Linear Programming" (*IBM Journal of Research and Development,* vol. 4, no. 5, 1960, pp. 507–517).

The existence of the method created a problem for the writer: The organization of the *STRATEGYST* was dictated by the methods of solution used. Ideally, the book should be entirely rewritten, but I was not prepared to do that. On the other hand, it still had much to offer the novice, it seemed to me, so I decided simply to put the new material in a new chapter, Chapter 6, at the end, and to hope that the reader would not be too upset by the organizational blemish. When he reaches the larger games—certainly when he reaches 4 × 4 games —he should turn to Chapter 6 and become familiar with the method set forth there.

I am again indebted to my associates for aid and education. M. E. Dresher, L. S. Shapley, A. W. Tucker, and P. Wolfe have read the new material critically and brightened dark corners of my mind in discussions of the pivot process.

I did not acknowledge in the Preface to the First Edition the tactful editorial assistance of Dorothy Stewart because she did her work after I had finished mine. I do so now, and repeat it for the present edition. Patricia Renfer has patiently prepared the manuscript, no matter how often I have torn it up.

The history of the book since its publication in 1954 may be of interest to a few of its readers. There has been a modest but persistent call for it—pleasing to the writer and astonishing, I believe, to the publisher —met by ten printings. Though not designed as a textbook, it is sometimes used as one—currently in nine colleges and universities, and it is a supplementary reading item in others. A French edition appeared in 1956, a Swedish edition in 1957, and a Russian edition in 1960. The publisher has arranged for Czech, Dutch, Japanese, Polish, and Spanish versions as well.

Some early printings were marred by an error, which came to light when the Swedish translator, Bjorn V. Tell, invited me to explain Example 13; I soon decided it would be easier to construct a new example than to construct a world in which the old one would look reasonable. I am grateful to Mr. Tell. The translator of the French edition, Mme. Mesnage, deleted from Example 12 my lighthearted speculation that Russian roulette might be a Good Thing as the Party sport. The Russian translator, Golubev-Novozhilov, renamed it American roulette, changed my cast of characters from Muscovite guards to United States Air Force types, and expressed the hope that I could tolerate having the wit turned against me. I sense at least one kindred spirit behind the Iron Curtain.

J. D. Williams

Preface to the First Edition

This book was conceived during a discussion among a group of persons who have been concerned for some years with the dual problems of the development and application of the Theory of Games of Strategy. While their immediate and primary interest has been applications pertinent to military affairs—particularly to those of the Air Force—their tangential interests in these problems are broad, practically without limit.

The sense of the discussion was that the activity, Game Theory, would benefit from having more persons informed regarding its nature; and that the knowledge would benefit the persons, of course. At the present time, this knowledge is mostly held by the tight professional group which has been developing the subject. Another, and larger, group has heard of it and comprehends, often dimly, its scope and character; the members of this group must usually accept, or reject, the ideas on the basis of insufficient knowledge. So it was felt to be worth while to try to bridge the gap between the priestly mathematical activity of the professional scientist and the necessarily blind reaction of the intelligent layman who happens not to have acquired a mathematical vocabulary.

Not recognizing that the discussion had reached a reasonable and natural stopping point, the group went on to nominate someone to write the book. After they coursed far and wide, and discussed many fine, though oblivious, candidates, it was anticlimactic to find myself chosen. The qualifications which won, or lost, the election are possibly worth enumerating: (1) I was at hand, and available; it is always immediately evident to research workers that an administrator is available. (2) While associated for some years, sympathetically, with the field of Game Theory, I was a complete ignoramus regarding most of its highly technical aspects; and I would probably not learn enough of these while writing the book to contaminate seriously the message that should be transmitted. (3) I was admirably situated, both organizationally and because of a natural bent to conserve energy, to call freely on my colleagues for aid and counsel; and the thus-shared burden would be more tolerable to all.

My intentions regarding the character of the book—its scope and style, and hence its audience—have changed several times during its production. In the end, it seems that we have produced a primer on Game Theory, for home study. We believe that you can sit down with it and, with the normal difficulties which attend intellectual effort—painful, but not mortal—learn to formulate and solve simple problems according to the principles of Game Theory. This, it seems to us, is itself a somewhat desirable good. The examples, it is hoped, will have a more important effect, for they are intended to touch the imagination; by couching them in terms of a very diverse set of activities, we have tried to encourage you to view some of yours in the light of Game Theory. Other than that, not much can be said for them, as they are frequently thin as to content and somewhat irreverent in tone. Actually, it is difficult to make them simple if not thin; and the tone is at once our insurance against the expert in the field of the example, who knows it is thin, and a device to keep the reader awake.

It turns out that there is nothing here of mathematics beyond arithmetic—extended to encompass negative numbers. That is to say, one needs to know how to add, subtract, multiply, and divide, using positive and negative numbers; usually whole numbers, however. The symbolism of mathematics is gone—a maddening restriction, we found, though self-imposed. Even so, there should be no illusions: Not everybody—even among persons with the desire—is going to be able to formulate and solve difficult problems in this field, for after all it is a very technical subject. But many can learn to handle some problems, and many more can appreciate the possibilities and limitations of the field.

We believe it possible that Game Theory, as it develops—or something like it—may become an important concept and force in many phases of life. To convert this from a possible to a likely event, several things must happen. Of these, one is that the potential user get in and use what is now available, or provide guidance (by suggesting problems) for the development of the theory so that it may eventually be useful to him.

Acknowledgments are very difficult to make, in a satisfactory way. I have shamelessly imposed on all my friends, and particularly on my colleagues, for ideas regarding examples. I have frequently taken these ideas, transposed them to strange settings, and used names similar (in fact, identical) to those of my benefactors to identify the actors. This made a dull writing job interesting; so they all aided me twice over.

However, fear of errors of omission—which are inevitable when one has so many creditors—cannot keep me from trying to record my debts. First of all, extensive manuscripts of M. E. Dresher, of RAND, and of the late J. C. C. McKinsey, formerly of RAND and then of Stanford University, which summarize many scores of papers by our colleagues and others, have been indispensable to my education. I am grateful for specific ideas to R. E. Bellman, R. L. Belzer, N. C. Dalkey, J. M. Danskin, W. H. Fleming, D. R. Fulkerson, Melvin Hausner, Olaf Helmer, S. M. Johnson, Abraham Kaplan, A. M. Mood, E. W. Paxson, E. S. Quade, R. D. Specht, and J. G. Wendel, all of RAND; and also to E. W. Barankin of the University of California, D. H. Blackwell of Howard University, S. C. Kleene of the University of Wisconsin, Oskar Morgenstern and A. W. Tucker of Princeton University, P. M. Morse of the Massachusetts Institute of Technology, R. M. Thrall of the University of Michigan, and C. B. Tompkins of George Washington University.

The following, in addition to some of those mentioned above, have read and commented on the manuscript, many in great detail: Bernard Brodie, G. W. Brown, R. W. Clewett, F. R. Collbohm, G. B. Dantzig, M. M. Flood, O. A. Gross, C. Hastings, Jr., B. W. Haydon, S. P. Jeffries, J. L. Kennedy, J. A. Kershaw, J. S. King, Jr., and R. A. Wagner, all of RAND; and E. F. Beckenbach of the University of California, Major General G. R. Cook, U.S. Army (ret.), C. F. Mosteller of Harvard University, and (probably) others. Dr. Dresher and Miss Wagner very kindly prepared the index. Miss Ruth Burns took the manuscript through several typed and vellum editions with improbable speed and accuracy.

I owe a very special debt to Warren Weaver of the Rockefeller Foundation, who was driven, by friendship and by interest in the topic, to read it very hard. His single-minded insistence on clarity of exposition was always of great value and sometimes a nuisance—especially in instances where his style and skill were better suited than my own to the problem of achieving clarity.

The RAND Corporation permitted me to rearrange my duties, which made it possible to write the book.

It is probably self-evident that when the author of so small a book owes so much to so many, the only things he can truly claim as his own are its faults. In this situation the 'compleat strategyst' would immediately form a coalition with Nature, and share these with Her.

J. D. Williams

Contents

CONTENTS

CHAPTER 3. THREE-STRATEGY GAMES

CHAPTER 4. FOUR-STRATEGY GAMES AND LARGER ONES

CHAPTER 5. MISCELLANY

The Compleat Strategyst

CHAPTER 1

Introduction

NATURE OF THE SUBJECT

It is all too clear at this moment that there are many ways for a book to begin; and most of those in plain sight are transparently bad. We are tantalized by the thought that somewhere among them *may* lie hidden a few having such noble qualities as these: The readers are informed—perhaps without suspecting it, though in the clearest prose—of what the writer intends to discuss; yet at the same time, it sounds like the Lorelei calling. Whereupon these readers resolve into two groups: The first, a large and happy family really, will stick to the book to the end, even though unimagined adversities impend. Further, this group will always think and speak kindly of it, and will doubtless have at least one copy in every room. The second group is most briefly described by stating that it differs from the first; but the book acts immediately as a soporific on all unpleasant passions, so, as it is sleepily laid aside, the sole lasting impression is that of a good gift suggestion.

If we could devise an opening strategy such as that, it would wonderfully exemplify the theme and aims of the book, for *our concern throughout will be with a method for selecting best strategies,* even in contexts where the word 'strategy' itself may not be in common use.

The contexts of interest to us are those in which people are at cross-purposes: in short, conflict situations. The problem of how to begin this book is recognizably of that type, for certainly you and the writer are at cross-purposes, as our interests are opposed—in a polite way, of course, but definitely opposed. For we hope to cozen you into a very difficult type of intellectual activity, while you, a reasonable person with enough troubles already, may crave only relaxation or satisfaction of curiosity. This conflict of interests is essential in the situations we shall study.

Another element is also essential and it is present here too: Each of us can exert *some* control over the situation. Many ways will occur to you: for one, you may throw the book at the cat, thus irritating both the writer and the cat,

but at some cost in property, perhaps some in self-respect, and undoubtedly some in deteriorated relations with the cat. Or you may skim the hard parts, and so on. There are aspects within the control of the writer, too, such as the choice and treatment of content—but it is not necessary to labor the point. And a further characteristic element appears: Some aspects of the situation are not within the control of either of us; for example, a multitude of events in our pasts and extraneous influences during the writing and reading periods will play important roles. Of course this particular problem, of beginning the book in a really optimum way, has a further characteristic which we shall henceforth shun, namely, it is too hard—else we should have solved it.

The restrictions on the subject matter being so few and mild, it follows that the set of conflict situations we are willing to consider is most notable for its catholicity. There is no objection, in principle, to considering an H-bomb contest between Mars and Earth, or a love affair of the Barrett-Browning type. The contest may be economic in character, or it may be Musical Chairs. Or it may be almost any one of the myriad activities which take place during conventional war. It doesn't follow that we have a nostrum for strategic ills in all these fields, but there is a possibility that our offering may as a method, perform useful service in any of them.

The method which will be presented is identified by the catch phrase *Game Theory* or, time permitting, the *Theory of Games of Strategy*. If this is your first encounter with that unlikely sequence of nouns, the sole reaction is probably: Why? Well, the idea takes its name from the circumstance that the study of games is a useful and usable starting point in the study of strategy. That does not really help, for again we hear: Why? Well, because games contain many of the ingredients common to all conflicts, and they are relatively amenable to description and to study. (Incidentally, having used the word 'game' to name the theory, we then call any conflict a game when we are considering it in the light of the theory.)

To illustrate the point, let us run our minds over a Poker game, keeping watch for items which are significant in, say, a military conflict. You and four others are thus studying human nature, under a system of rewards, you hope. We note at once that the players have opposing interests; each wants to win and, because the winnings of one are necessarily the losses of another, their interests are opposed. This provides the basis of conflict. We observe too that some elements of the action, being personal choices, are completely within your control. And the same being true for each player, there are elements which are not within your control; worse, they are controlled by minds having interests inimical to yours. Finally, there are elements of the game that are not, under the rules, within the control of any player, such as the order of the cards in the deck. These elements may be thought of as being controlled by Nature—who has a massively stable personality, a somewhat puckish attitude toward your important affairs, but who bears you no conscious malice. These are all surely familiar aspects of any conflict situation.

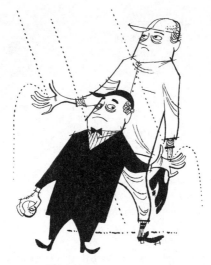

Another characteristic is that the state of information—intelligence, in the military sense—is a factor, and, as usual, is an imperfect and hence troublesome factor: We don't know what the other fellow's hole card is. There is also the bluff by which you, or the opposition, give false evidence regarding intentions or strength of forces. Other similarities will occur to you; people even get killed, occasionally.

But the analogy should not be pushed too far. You can think of many aspects of warfare which are not reflected in Poker. One tank will sometimes kill two tanks, in a showdown; whereas a pair of Jacks always wins over an Ace-high hand in the showdown. Of course Poker could be modified to make it contain showdown possibilities of this kind, say by ruling that an Ace is superior to any pair, up to Jacks, whenever anybody's wife phones during the play of a hand. But the fact is that games don't exhibit all the complexities of warfare and of other real-life conflict situations—which is precisely why they are usable starting points for a study of strategy. In the early stages of developing a theory it just is not possible simultaneously to handle very many interacting factors.

It is probably clear, then, that games do contain some of the basic elements that are present in almost any interesting conflict situation. Does it follow that we can learn useful things by beginning a study with them? Not necessarily. It may be that military, economic, and social situations are just basically too complicated to be approached through game concepts. This possibility gains credence from the fact that the body of Game doctrine now in existence is not even able to cope with full-blown real games; rather, we are restricted at present to very simple real games, and to watered-down versions of complicated ones, such as Poker.

It may be baffling then that someone devotes valuable energy to the study and development of Game Theory—and, moreover, expects you to participate! The reason it is done is in part an act of hope and of faith, stemming from past successes. For the invention of deliberately oversimplified theories is one of the major techniques of science, particularly of the 'exact' sciences, which make extensive use of mathematical analysis. If the biophysicist can usefully employ simplified models of the cell and the cosmologist simplified models of the universe, then we can reasonably expect that simplified games may prove to be useful models for more complicated conflicts.

Of course the mortality among such theories is higher than any military organization would tolerate in *its* activities, and those that are successful are not really immortal; the best that can be expected of one is that it be adequate for certain limited purposes, and for its day.

AN HISTORICAL THEORY

It may be useful to examine one successful scientific abstraction, to see what it is like and for the sake of the hints it may give us. We choose one which is surely an example of heroic oversimplification.

Let us assume that we may, in order to study their motions, replace each of the major bodies of the Solar System by a point; that each point has a mass equal to that of the body it replaces; that each pair of points experiences a mutual attraction; that we may estimate the attractive force by multiplying the mass of one point by the mass of the other, after which we divide that product by the square of the distance between the points; that we may neglect all else; and that it isn't patently stupid to consider this theory, else we would never get started.

The fact is that this theory, the Theory of Gravitation, has been adequate for predicting the motions of the planets for two and one-half centuries—and this in the face of constant checking by positional astronomers, who, it can fairly be said, carry precision to extremes. The worst strain has come from the orbit of Mercury, which unaccountably drifted from the predicted place by one-fifth of a mil (a foot, at a distance of a mile) *per century,* thus showing that the theory is rough after all, just as it looks. The improved theory, by Einstein, accounts for this discordance.

LESSONS AND PARALLELS

The elements of the theory stated above of course did not just float into a mind dazed by a blow from an apple. There was much information at hand regarding the actual behavior of the planets, thanks largely to Tycho Brahe, and a wearisome mess it was. Kepler finally grubbed out of it a few rules of thumb; with these, and with a lift from a new mathematical invention (the Calculus), Newton soon afterward hit upon the above abstraction. He had the misfortune to try it immediately on the Moon, which cost him years of happiness with his theory, for the data were seriously in error.

This example contains several lessons for us. One is that theories may be very simple, while the phenomena they model do not appear simple. Anybody who supposes that planetary motions are quite simple has never had the responsibility for predicting them; the ancients had good reason to name them the Wanderers. Another lesson is that a theory can be very general, being applicable to a wide variety of phenomena, without being sterile; the Theory of Gravitation is even more general than stated above, for it applies to *all* mass particles, not just to the major bodies of the Solar System. Another lesson is that theories often or usually are imperfect, though the one used as an example is embarrassingly good. Another—and this is a very important one—is that the theory covers only one of the interesting factors which may affect the motion of bodies; one, moreover, that is frequently negligible. For example, the gravitational attraction between two airplanes flying a tight formation is equivalent to the weight of a cigarette ash, perhaps a sixteenth of an inch long.

Still another lesson concerns the importance of having some relevant data. In this respect Newton was somewhat better off than we are—we who are trying to do abstraction in such a field as conflict. For most of the data we have on man relate to the individual—his physical and mental composition, health, ability, etc.—and, to a lesser extent, to the gross characteristics of the social group. The *interactions* between men, as individuals in a group or between groups, have not been studied on anything approaching the scale needed; and these interactions are the stuff which constitutes conflict.

Another lesson, or at least a suggestive note, is the fact that Newton almost simultaneously developed the Theory of Gravitation and a new branch of mathematics—the Calculus; and the theory would have been practically unusable without it. In fact the Calculus has played a dominant role in all physical science for a quarter of a millennium. It is provocative to speculate on whether Game Theory will develop a new mathematical discipline destined for a comparable role in analyzing the interactions of men. It is much too early to conjecture that it will; so far, there has been little that is recognizable as brand new, and much that is recognizable as borrowings from established branches. But it may happen, and perhaps even it must happen if the application of the method is to reach full flower. It is at least interesting that the original development of Game Theory is the work of one of the really great mathematicians and versatile minds of our day—John von Neumann.*

Game Theory is very similar in spirit to the Theory of Gravitation. Both attempt to treat broad classes of events according to abstract models. Neither tries to model all the complexities present in any situation. One of them, to the extent it is applicable to animal activity, concerns itself with some of the involuntary actions; thus the Theory of Gravitation can answer superbly all questions regarding the gross motions of a pilot, alone at 40,000 feet, who is unencumbered by aircraft, parachute, or other device. Game Theory, on the other hand, would be more interested in the strategy by which he achieved all this and with questions regarding its optimality among alternative strategies; it, therefore, enters the region of decisions and free will.

This comparison with Gravitation Theory will be unfortunate if it seems to imply comparable utility and (in a loose sense) validity—not to say social standing—for the two theories. The one is mature and comfortably established as a useful approximation to Nature, whereas the other is a lusty infant, which may be taken by a plague or which may grow up to great importance, but which is now capable only of scattered contributions. As an infant, it is proper for it to be a little noisy.

Having permitted you to sense the galling bit of mathematics that will come (i.e., 'bit' as in the horse), we hasten to assure you that the approach we shall use is that of the primer, strictly, which means (you will recall) an elementary book for practice in spelling, and the like. We assume explicitly that you are not trained in mathematics beyond rudimentary arithmetic. In fact, if this is not true, simple charity requires that you close the book.

SECTARIAN REMARKS ON METHOD

It is sometimes felt that when phenomena include men, it is tremendously more difficult to theorize successfully; and our relative backwardness in these matters seems to confirm this. However, part

---

* Von Neumann's first paper on Game Theory was published in 1928, but the first extensive account appeared in 1944: *Theory of Games and Economic Behavior* by John von Neumann and Oskar Morgenstern (Princeton University Press, Princeton, N. J.). The challenging nature of this work was immediately appreciated by some reviewers, such as A. H. Copeland, who wrote "Posterity may regard this book as one of the major scientific achievements of the first half of the twentieth century" (*Bulletin of the American Mathematical Society,* vol. 51, 1945, pp. 498–504).

of the so-far minor effort made in this direction has been dissipated against hand-wringing protestations that it is too hard to do. Some of the impetus toward simple theory—simple theory being a few axioms and a few rules for operating on them, the whole being more or less quantitative—has come from amateurs; physical scientists, usually. These are often viewed by the professional students of man as precocious children who, not appreciating the true complexity of man and his works, wander in in wide-eyed innocence, expecting that their toy weapons will slay live dragons just as well as they did inanimate ones. Since Game Theorists are obviously children of this ilk, you doubtless anticipate that we shall now make some reassuring sounds, probably at the expense of the professionals, else we should not have raised the subject. If you do so anticipate, this shows how easy it really is, for it establishes you as a promising student of man, too!

The motive force that propels the Game Theorist isn't *necessarily* his ignorance of the true complexity of man-involved conflict situations; for he would almost surely try to theorize if he were not so ignorant. We believe, rather, that his confidence—better, his temerity—stems from the knowledge that he and his methods were completely outclassed by the problems of the inanimate world. He could not begin to comprehend them when he looked at them microscopically and, simultaneously, with a wide field of view; the quantity of detail and the complexity of its organization were overpowering. So, since he has had some success in that field, he suspects that sheer quantity and complexity cannot completely vitiate his techniques.

He is also aware that his successes occur spottily, so that his knowledge is much less complete than the uninitiated suspect—the uninitiated including of course those who believe the animate field must be vastly harder than the inanimate *because* the latter has been done so well (!). For example, modern physicists have only the foggiest notions about some atomic constituents—though they designed successful A-bombs. Their favorite particle, the electron, is shrouded in ignorance; such elementary information as where-is-it and, simultaneously, where-is-it-going is not known—worse, they have decided this information is in a strict sense forever unknowable. The mathematicians are likewise a puny breed. Item: after centuries of effort, they still don't know the minimum number of colors needed to paint a map (so that adjacent countries will not have the same color); it's fair to add that they suspect the number is four, but they haven't proved it.

Within the last hundred years, the physical scientists have added a

new arrow to their quiver, one which has played only the role of minor weapon in most of their campaigns so far, namely, mathematical statistics. They are now beginning to find more important uses for it, and there is a good prospect that it will become an increasingly important tool in the animate field; Game Theory has many points of contact with it. An early demonstration of its power, and a harbinger of its range of utility, was its success in accounting for the distribution (over the years) of deaths in Prussian Army Corps due to kicks from horses.* If you protest that horses are more predictable than men, we counter confidently with the assertion that the method is just as applicable to the distribution of horses kicked to death by Prussians. Of course the

---

* The reference is to these data, covering ten army corps over a twenty-year period (1875–1894). The deaths are per corps, per year.

Deaths	Occurrences Observed	Occurrences Computed
0	109	109
1	65	66
2	22	20
3	3	4
4	1	1

The computed values are derived from one bit of observed information, namely, that the fatalities average about one every twenty months, and from a statistical theory that is particularly applicable to rare events.

whole field of insurance is an example of statistical theory applied to some aspects of human affairs; the balance sheets of the insurance companies bear eloquent testimony to its success. Humans are not completely unpredictable.

So what are reasonable expectations for us to hold regarding Game Theory? It is certainly much too simple a theory to blanket all aspects of interest in any military, economic, or social situation. On the other hand, it is sufficiently general to justify the expectation that it will illumine certain critical aspects of many interesting conflict situations.

There are at present some important things to be done. One is to develop further the theory itself, so that more difficult and more varied problems can be solved; this task falls to the scientists. Another is to find situations to which existing theory can profitably be applied; one purpose of this book is to increase the number of persons who, by knowing the rudiments of the theory, can suggest applications to problems selected from those they encounter. (Those who hang on far enough will be able to formulate and solve simple problems for themselves.) Another task is the collection of data in the field of human interaction, to improve the bases of abstraction.

PLAYERS AND PERSONS

Now to Game Theory itself. We shall begin by looking at Stud Poker, and we shall look just long enough to introduce some concepts and terminology that will be used throughout the book. You and four others are still sitting there, with a deck of cards, some money or other valuables, and an agreed-on set of rules that covers all contingencies. The rules govern how the cards are to be doled out, who may bet and when, how the various hands are to be judged in the showdown, and what happens to the pot.

One of the obvious things about this situation is that it is a five-person game. But this may be more obvious than true; for perhaps two of the players formed a coalition, in advance of the game, in which they agreed to pool their winnings or losses. If they did so, it is reasonable to suppose that they will play for their common good whenever circumstances permit it. Thus if one member of the coalition believes his partner has a good chance of winning a particular hand, he should take whatever action he can toward the common good. If only three hands are active, perhaps he should fold so that the burden of calling falls on the outsider, or perhaps he should raise the bet in order to increase the pot, even though he knows his cards cannot win the hand. In short, the members of the coalition will behave as much like a single individual, with two heads, as they can.

In the case where two players have formed a coalition, it is evident that it may be fruitful to consider it as a four-person, rather than as a five-person, game. Thus we come to believe *it is significant to count the number of sets of opposing interests* around the table, rather than the bodies. According to this principle, Bridge is classed as a two-person game, because there are only two sets of interests when the partners are permanent. You will note that the words 'person' and 'player,' as we use them, cover legal persons and organizations, as well as natural persons.

Again, you may prefer to regard the Poker game as a two-person game in which you are one of the players and the other four individuals are the other player. If they do not look at it the way you do, they will gain no advantage from the association you have imagined for them, and you will suffer no loss from it; it is as though they constitute a coalition with weak internal communications, or some other malady which makes it ineffective.

This is one of the fundamental distinctions in Game Theory, namely, the number of persons—distinct sets of interests—that are present in the game. The form of analysis and the entire character of the situation depend on this number. There are three values, for the number of persons, which have special significance: one, two, and more-than-two.

Solitaire is an example of a one-person game when played for recreation, for your interests are the only ones present. Even if you buy the deck for, say, $1 a card from somebody who is willing to pay you, perhaps, $5 a card for all cards transferred to the payoff piles, the case is the same: only chance events must be countered, and not the moves of a responsive human adversary. One-person games are uninteresting, from the Game Theory point of view, and therefore are not really studied here. Their solution is quite straightforward, conceptually: You simply select the course of action that yields the most and do it. If there are chance elements, you select the action which yields the most on the average, and do it. You may complain that we are glossing over an awful lot of practical difficulties; and that's right.

However, one-person games (including Solitaire) may be regarded as a special kind of two-person game in which you are one of the players and Nature is the other. This may be a useful viewpoint even if you don't believe that Nature is a malignant Being who seeks to undo you. For example, you may not know enough about Nature's habits to select the course which will yield the most on the average. Or it may happen that you know the kinds of behavior open to Nature, but know little about the frequency with which She uses them. In this case Game Theory does have something to say; it will lead you to conservative play, as we shall see later.

The true two-person game is very interesting. It occurs frequently and its solution is often within our present means, both conceptual and technological. This is the common conflict situation. You have an opponent who, you must assume, is intelligent and trying to undo you. If you choose a course of action which appears favorable, he may discover your plans and set a trap which capitalizes on the particular choice you have made. Many situations which are not strictly two-person games may be treated as if they were; the five-man Poker game was an example of this, where you could assign the interests present at the table to two 'persons,' yourself and everybody-not-you. Most of the work done to date in Game Theory deals with the two-person game.

When the number of distinct persons, i.e., sets of interests, exceeds two, qualitatively new things enter. The principal new factor is that the identities of the persons may change in the course of the game, as temporary coalitions are formed and broken; or certain players may form what is in effect a permanent partial coalition in some area of action where they conceive it to be beneficial. This could happen in the Poker game and would compromise our treatment of it as a two-person game, as proposed earlier. For example, you might wish to team up with others, informally but effectively, to act against a heavy winner; you might be motivated by fear that he would leave the game taking most of the cash with him, or you might prefer to see more of it in the hands of a weaker player. Our under-standing of games that involve more than two persons is less complete at present than for two-person games, and the subject is rather compli-cated; in fact, it lies beyond the modest limits of this book.

THE PAYOFF

We have indicated that the number of persons involved is one of the important criteria for classifying and studying games, 'person' meaning a distinct set of interests. Another criterion has to do with the payoff: What happens at the end of the game? Say at the end of the hand in Poker? Well, in Poker there is usually just an exchange of assets. If there are two persons, say you (Blue) and we (Red), then if you should win $10, we would lose $10. In other words,

$$\text{Blue winnings} = \text{Red losses}$$

or, stated otherwise,

$$\text{Blue winnings} - \text{Red losses} = 0$$

We may also write it as

$$\text{Blue payoff} + \text{Red payoff} = \$10 - \$10 = 0$$

by adopting the convention that winnings are positive numbers and that losses are negative numbers.

It needn't have turned out just that way; i.e., that the sum of the payoffs is zero. For instance, if the person who wins the pot has to contribute 10 per cent toward the drinks and other incidentals, as to the cop on the corner, then the sum of the payoffs is not zero; in fact

$$\text{Blue payoff} + \text{Red payoff} = \$9 - \$10 = -\$1$$

The above two cases illustrate a fundamental distinction among games: It is important to know whether or not the sum of the payoffs, counting winnings as positive and losses as negative, to all players is zero. If it is, the game is known as a *zero-sum game*. If it is not, the game is known (mathematicians are not very imaginative at times) as a *non-zero-sum game*. The importance of the distinction is easy to see: In the zero-sum case, we are dealing with a good, clean, closed system; the two players and the valuables are locked in the room. It will require a certain effort to specify and to analyze such a game. On the other hand, the non-zero-sum game contains all the difficulties of the zero-sum game, plus additional troubles due to the need to incorporate new factors. This can be appreciated by noting that we can restore the situation by adding a fictitious player—Nature again, say, or the cop. Then we have

$$\text{Blue payoff} = \$9$$
$$\text{Red payoff} = -\$10$$
$$\text{Cop payoff} = \$1$$

so now

$$\text{Blue payoff} + \text{Red payoff} + \text{Cop payoff} = \$9 - \$10 + \$1 = 0$$

which is a *three-person zero-sum* game, of sorts, where the third player has some of the characteristics of a millstone around the neck. But recall that we don't like three-person games so well as we do two-person games, because they contain the vagaries of coalitions. So non-zero-sum games offer real difficulties not present in zero-sum games, particularly if the latter are two-person games.

Parlor games, such as Poker, Bridge, and Chess, are usually zero-sum games, and many other conflict situations may be treated as if they were. Most of the development of Game Theory to date has been on this type of game. Some work on non-zero-sum games has been done, and more is in progress, but the subject is beyond our scope. A troublesome case of particular interest is the two-person game in which the nominally equal payoffs differ in utility to the players; this situation occurs often even in parlor games.

STRATEGIES

Just as the word 'person' has a meaning in Game Theory somewhat different from everyday usage, the word 'strategy' does too. This word, as used in its everyday sense, carries the connotation of a particularly skillful or adroit plan, whereas in Game Theory it designates any *complete* plan. *A strategy is a plan so complete that it cannot be upset by enemy action or Nature;* for everything that the enemy or Nature may choose to do, together with a set of possible actions for yourself, is just part of the description of the strategy.

So the strategy of Game Theory differs in two important respects from the conventional meaning: It must be utterly complete, and it may be utterly bad; for nothing is required of it except completeness. Thus, in Poker, all strategies must make provision for your being dealt a Royal Flush in Spades, and some of them will require that you fold instantly. The latter are not very glamorous strategies, but they are still strategies—after all, a Bridge player once bid 7 No-Trump while holding 13 Spades. In a game which is completely amenable to analysis, we are able—conceptually, if not actually—to foresee all eventualities and hence are able to catalogue all possible strategies.

We are now able to mention still another criterion according to which games may be classified for study, namely, the number of strategies available to each player. Thus, if Blue and Red are the players,

Blue may have three strategies and Red may have five; this would be called a 3×5 game (read 'three-by-five game').

When the number of players was discussed, you will recall that certain numbers—namely, one, two, and more-than-two—were especially significant. Similarly, there are critical values in the number of strategies; and it turns out to be important to distinguish two major categories. In the first are games in which the player having the *greatest* number of strategies still has a finite number; this means that he can count them, and finish the task within some time limit. The second major category is that in which at least one player has infinitely many strategies, or, if the word 'infinitely' disturbs you, in which at least one player has a number of strategies which is larger than any definite number you can name. (This, incidentally, is just precisely what 'infinitely large' means to a mathematician.)

While infinite games (as the latter are called) cover many interesting and useful applications, the theory of such games is difficult. 'Difficult' here means that there are at least some problems the mathematician doesn't know how to solve, and further that we don't know how to present any of it within the friendly pedagogical limits of this book; such games require mathematics at the level of the Calculus and beyond— mostly beyond. Therefore we here resolve to confine our attention to finite games.

We shall find it convenient, in later chapters, to distinguish three cases among finite games: namely, those in which the player having the *least* number of strategies has exactly two, exactly three, or more-than-three. In addition, considerations of labor, fatigue, and the better life will cause us to develop a rather special attitude toward games having more than about ten strategies.

THE GAME MATRIX

We are now in a position to complete the description of games, i.e., conflict situations, in the form required for Game Theory analysis. We will freely invoke all the restrictions developed so far, so as to aim the description directly at the class of games which will be studied throughout the book. Hence our remarks will primarily apply to finite, zero-sum, two-person games.

The players are Blue and Red. Each has several potential strategies which we assume are known; let them be numbered just for identification. Blue's strategies will then bear names, such as Blue 1, Blue 2, and so on; perhaps, in a specific case, up to Blue 9; and Red's might range from Red 1 through Red 5. We would call this a nine-by-five game and write it as '9 × 5 game.' Just to demonstrate that it is possible to have a 9 × 5 game, we shall state one (or enough of it to make the point). Consider a game played on this road map:

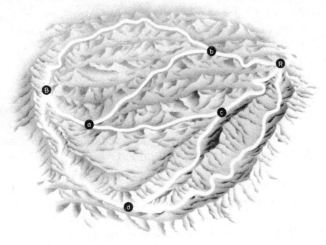

The rules require that Blue travel from *B* to *R*, along the above system of roads, without returning to *B* or using the same segment twice during the trip. The rules are different for Red, who must travel from *R* to *B*, always moving toward the west. Perhaps Blue doesn't want to meet Red, and has fewer inhibitions about behavior. You may verify that there are nine routes for Blue and five for Red.*

---

* To avoid even the possibility of frustrating you this early in the game, we itemize the routes. Blue may visit any of the following sets of road junctions (beginning with *B* and ending with *R* in each case):

b, bac, bacd, ab, ac, acd, dcab, dc, d

Red may visit

b, ba, ca, cd, d

The rules must also contain information from which we can determine what happens at the end of any play of the game: What is the payoff when, say, Blue uses the strategy Blue 7 (the northern route, perhaps) and Red uses Red 3 (the southern route, perhaps)? There will be $9 \times 5 = 45$ of these pairs and hence that number of possible values for the payoff; and these must be known. Whatever the values are, it is surely possible to arrange the information on this kind of bookkeeping form:

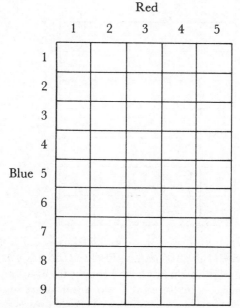

Such an array of boxes, each containing a payoff number, is called a *game matrix*. We shall adopt the convention that a positive number in the matrix represents a gain for Blue and hence a loss for Red, and vice versa. Thus if two of the values in the game matrix are 3 and −8, as shown here,

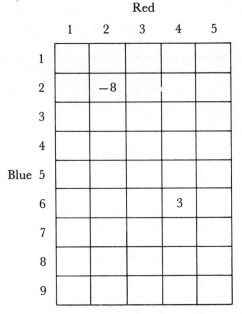

the meaning is: When Blue uses Blue 6 and Red uses Red 4, Blue wins 3 units, whereas when Blue 2 is used vs. Red 2, Red wins 8 units.

When the original problem has been brought to this form, a Game Theory analysis may begin, for all the relevant information is represented in the descriptions of the strategies whose signatures border the matrix and in the payoff boxes. This is the Game Theory model of the conflict, and the applicability of the subsequent analysis will depend completely on the adequacy of this form of representation—a set of strategies and a payoff matrix.

IMPLICIT ASSUMPTIONS

Perhaps the last statement should be expanded. We narrow our attention for a moment to two complicated objects: One is the real conflict situation in which Blue and Red are involved. This includes the rules, regulations, taboos, or whatnots that are *really* operative; it includes the true motives of the players, the geography, and in fact everything that is significant to the actual game. The second object is also real, but it is much more simple: It is the rules we have *written,* the strategies we have enumerated and described *on paper,* and the game matrix we have *written.* There is a relationship—a significant one, we trust—between these two objects. The second object—the marks on paper—is an abstraction from the first. We can discover some non-obvious properties of this second object by making a Game Theory analysis, and these properties *may* have some validity in connection with the first object—the real world game. It will all depend on the adequacy of the abstraction.

The principal topic of this book will be discussion of how Game Theory operates on the second object, the abstract model. The difficulties and questions that will come up in that discussion will be, principally, technical ones, rather than conceptual ones. They will be questions of ingenuity in handling difficult mathematical problems or of devices to avoid outrageous labor; in general, just high-class crank turning. We should recognize, before passing to this relatively comfortable pastime, that all is now easy because we have already glided over many of the real difficulties, namely, the conceptual ones.

One of the conceptual problems, a critical point in Game Theory so far as its application to real-life conflict situations is concerned, is reached when we try to fill in the boxes with the values of the payoff. While there will be individual cases in which the requirements are less severe, in general we have to assume that the payoff can, in principle, be measured numerically; that we in fact know how to measure it; and that we do measure it, with sufficient accuracy. Further, the units of measurement must be the same in all boxes, and the units must be simple, dimensionally; that is to say, we are not prepared to cope with dollars in one box, grams of uranium in another, and lives in another—unless of course we happen to know exchange ratios between these items and can eliminate the heterogeneity of units of measurement. If the payoff in each box contains several numbers representing disparate items—which may be some dollars, some uranium, and some lives, say—we are still in trouble, except in special cases. This difficulty can arise in ordinary games; consider, for example, a two-person game between master players; the stakes may be sums of money and prestige. Unless we are prepared to adopt an exchange ratio between prestige and money, our analysis is likely to be in trouble.

Another conceptual difficulty in connection with real problems is that of defining the problem sufficiently crisply, so that the action alternatives available to the players may be completely itemized; and to do this without isolating the problem from the important influences of its original environment.

Other hazards will be pointed out from time to time in later chapters, as the discussion veers by an occasional rock. There will be some things to note on the other side of the question, too; for the model need not be an exact replica of the real-life situation in order to be useful. We shall see that there is sometimes considerable latitude in these matters.

THE CRITERION

A perennial difficulty in modelmaking of the analytical (as opposed to wooden) variety is the illness which might well be known as criterion-trouble. What is the criterion in terms of which the outcome of the game is judged? Or should be judged?

To illustrate the wealth of possible criteria in a homely example, consider a housewife who has $5 to spend on meat. What should she buy? If her criterion is simply quantity, she should buy the cheapest kind and measure the payoff in pounds. If it is variety, she should buy minimum, useful quantities of several kinds, beginning with the cheap-

est kinds; she measures the payoff by the number of kinds she buys. Or she may be interested in protein, fat, or calories. She may have to satisfy various side conditions, or work within certain constraints, such as allergies, tastes, or taboos. She may be interested in least total effort, in which case she may say, "I want five dollars worth of cooked meat—the nearest, of course—and deliver it sometime when you happen to be down our way."

Generally speaking, criterion-trouble is the problem of what to measure and how to base behavior on the measurements. Game Theory has nothing to say on the first topic, but it advocates a very explicit and definite behavior-pattern based on the measurements.

It takes the position that there is a definite way that rational people should behave, if they believe in the game matrix. The notion that there is some way people ought to behave does not refer to an obligation based on law or ethics. Rather it refers to a kind of mathematical morality, or at least frugality, which claims that the *sensible object of the player is to gain as much from the game as he can, safely, in the face of a skillful opponent who is pursuing an antithetical goal.* This is our model of rational behavior. As with all models, the shoe has to be tried on each time an application comes along to see whether the fit is tolerable; but it is well known in the Military Establishment, for instance, that a lot of ground can be covered in shoes that do not fit perfectly.

Let us follow up the consequences of this model in a zero-sum game, which, you will recall, is a closed system in which assets are merely passed back and forth between the players. It won't affect anything adversely (except Red), and it will simplify the discussion, if we assume for a moment that all payoffs in the game matrix are *positive;* this means that the strategy options available to the players only affect how many valuables Red must give to Blue at the end of a play of the game; this isn't a fair game for Red, but we will let him suffer for the common weal.

Now the viewpoint in Game Theory is that *Blue wishes to act in such a manner that the least number he can win is as great as possible, irrespective of what Red does;* this takes care of the safety requirement. *Red's comparable desire is to make the greatest number of valuables that he must relinquish as small as possible, irrespective of Blue's action.* This philosophy, if held by the players, is sufficient to specify their choices of strategy. If Blue departs from it, he does so at the risk of getting less than he might have received; and if Red departs from it, he may have to pay more than he could have settled for.

The above argument is the central one in Game Theory. There is a way to play every two-person game that will satisfy this criterion. However, as in the case of the housewife buying meat, it is not the only possible criterion; for example, by attributing to the enemy various degrees of ignorance or stupidity, one could devise many others. Since Game Theory does not attribute these attractive qualities to the enemy, it is a conservative theory.

You will note an apparent disparity in the aims of Blue and Red as stated above; Blue's aims are expressed in terms of winning and Red's in terms of losing. This difference is not a real one, as both have precisely the same philosophy. Rather, it is a consequence of our convention regarding the meaning of positive and negative numbers in the game matrix. The adoption of a uniform convention, to the effect that Blue is always the maximizing player and Red the minimizing player, will reduce technical confusion (once it becomes fixed in your mind); but let's not pay for this mnemonic by coming to believe that there is an essential lack of symmetry in the game treatment of Blue and Red.

EXAMPLE 1. THE CAMPERS

It may help to fix these ideas if we give a specific physical realization. When the payoffs are all positive, we may interpret them as the altitudes of points in a mountainous region. The various Blue and Red strategies then correspond to the latitudes and longitudes of these points.

To supply some actors and motivation for a game, let's suppose that a man and wife—being very specific always helps, so let's name them Ray and Dotty—are planning a camping trip, and that Ray likes high altitudes and Dotty likes low altitudes. The region of interest to them

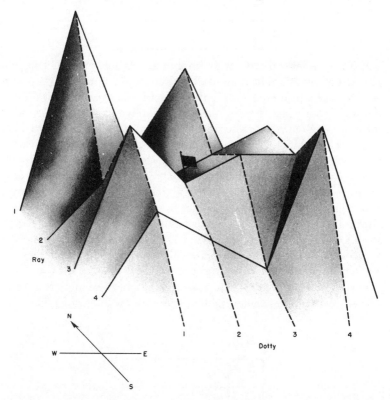

is crisscrossed by a network of fire divides, or roads, four running in each direction. The campers have agreed to camp at a road junction. They have further agreed that Ray will choose the east-west road and that Dotty will choose the north-south road, which jointly identify the junction. If Game Theory doesn't save them, frustration will kill them.

The junctions on the roads available to Ray have these altitudes (in thousands of feet):

1	7	2	5	1
2	2	2	3	4
Ray 3	5	3	4	4
4	3	2	1	6

Being a reasonable person, who simply wants to make as much as possible out of this affair, he is naturally attracted to the road Ray 1—with junctions at altitudes of 7, 2, 5, and 1—for it alone can get him the 7-thousand-foot peak. However, he immediately recognizes this kind of thinking as dream stuff; he does not dare undertake a plan which would realize him a great deal if it succeeds, but which would lead to disaster if Dotty is skillful in her choice. Not anticipating that she will choose carelessly, his own interests compel him to ignore the breathtaking peaks; instead, he must attend particularly to the sinks and lows, of one kind and another, which blemish the region. This study leads him finally to the road Ray 3, which has as attractive a low as the region affords, namely, one at an altitude of 3 thousand feet. By choosing Ray 3, he can ensure that the camp site will be *at least* 3 thousand feet up; it will be higher, if Dotty is a little careless.

His wife—as he feared—is just as bright about these matters as he is. The critical altitudes on her roads are listed in the following table:

Dotty

1	2	3	4
7	2	5	1
2	2	3	4
5	3	4	4
3	2	1	6

As she examines these, she knows better than to waste time mooning over the deep valleys of Dotty 3 and Dotty 4, much as she would like to camp there. Being a realist, she examines the peaks which occur on her roads, determined to choose a road which contains only little ones. She is thus led, finally, to Dotty 2, where a 3-thousand-foot camp site is the worst that can be inflicted on her.

We now note that something in the nature of a coincidence has occurred. Ray has a strategy (Ray 3) which guarantees that the camp site will have an altitude of 3 thousand feet or more, and Dotty has one (Dotty 2) which ensures that it will be 3 thousand feet or less. In other words, either player can get a 3-thousand-foot camp site by his own efforts, in the face of a skillful opponent; and he will do somewhat better than this if his opponent is careless.

When the guaranteed minimum and maximum payoffs of Blue and Red are exactly equal, as they are here, the game is said to have a *saddle-point*, and the players should use the strategies which correspond to it. If either alone departs from the saddle-point strategy, he will suffer unnecessary loss. If both depart from it, the situation becomes completely fluid and someone will suffer.

Note too this consequence of having a saddle-point: security measures are not strictly necessary. Either Ray or Dotty can openly announce a choice (if it is the proper one), and the other will be unable to exploit the information and force the other beyond the 3-thousand-foot site.

We remarked that the existence of a Game Theory saddle-point is something of a coincidence. Yet it corresponds to a pass or saddle-point in the mountains, and almost any complicated arrangement of mountains will contain many passes. The trick is that a mountain pass must have special features to make it qualify as a Game Theory saddle-point. For one, the road through the pass must run north and south; i.e., this road must lie within the choice of the player who wants to keep the payoffs small. Another feature is that there must be no high ground north or south of the pass. Another is that there must be no low ground east or west of the pass. It is rather reasonable to find qualifications such as these; for after all a mountain pass is a *local* feature of the terrain, so some additional qualities are needed to ensure that it have the global properties of being best over the entire region.

While we shall always find it worth while to inspect games for saddle-points, the incidence of saddle-points is not very great, in general. In a 4 × 4 game, such as the present one, there is about one chance in ten that a matrix of random numbers will have a saddle-point.

In the present instance, the saddle-point can be eliminated by making an apparently minor change in the matrix, at any one of several points. For instance, if the altitude of the junction at the intersection of Ray 4 with Dotty 2 were changed from 2 to 6, the character of the game would become very different. In that game, i.e., in

Dotty

		1	2	3	4
	1	7	2	5	1
	2	2	2	3	4
Ray	3	5	3	4	4
	4	3	6	1	6

our elementary deductions regarding choice of strategies break down. If Ray argues as before, he will be led again to the road Ray 3, which ensures that the came site will be 3 thousand feet up, or higher; but Dotty will be led to Dotty 3 this time, which only guarantees that the camp site will be at 5 thousand feet, or less.

Thus there is a gap, between 3 and 5 thousand feet, in which the sit-

uation is out of control. Your intuition may suggest that there should be a way to play the game which will close this gap. In fact there is a way; but we must begin our study with simpler situations. In passing, we remark that good play will now require a more elaborate security system than was needed in the case of a saddlepoint. In particular, the players will need to express their choices of stategy simultaneously, or in sealed ballots. What they should write on these ballots is quite a problem.

Two-strategy Games

PART ONE: 2×2-Games

THE APPROACH

We shall enter upon the technical ground of game analysis as gently as we can. The 2×2 game looks like the appropriate lightweight vehicle.

You will recall: we require that the scheme for our game be reduced to a payoff matrix—a rectangular array (possibly square) of numbers, indexed against the various strategies which are available to the players. Conversely, it is rather evident that *any* rectangular array of numbers may be thought of as the payoff matrix for some game; we can always invent a game which has that payoff matrix. We shall, therefore, attend at times to the properties and implications of specific matrixes, without worrying about the actual games to which they are pertinent. By thus eliminating the substantive material, we shall have an uncluttered view of the technical matters.

We shall always be seeking *solutions* to games. This means that we shall try to discover which strategy or strategies the players should use and, if more than one is required, how priorities should be assigned; for in any given instance, only one strategy may actually be used. Every game of the type we shall consider does have a solution; i.e., there is a good way to play it and we can find it, if the actual labor happens not to be prohibitive.

We begin by taking simple payoff matrixes, and shall later work our way into nonsimple ones. This chapter will emphasize 2×2 games, but it will not be confined to them; it will include 2×3 games, 2×4 games, and in fact any game in which *one* of the two players has but two strategies.

We shall always use these conventions: Blue's strategies are listed and indexed in a column along the left edge of the game matrix, Red's in a row along the top edge. The payoffs are to Blue; thus *a positive number indicates a payment to Blue from Red, a negative number a payment from Blue to Red.* The reader having a proper outlook will find it helpful to identify himself with Blue. Positive numbers then reasonably mean that he gets something, while negative numbers mean that he pays.

FLUCTUATIONS

Let us begin with an easy one:

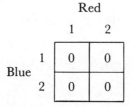

You will recall the meaning of this diagram: the integers, 1 and 2, standing at the left of the set of boxes, are simply the *names* of Blue's strategies—the two courses of action available in this game to Blue. Similarly, the two integers on top stand for Red Strategy 1 and Red Strategy 2. The numbers in the boxes tell us the payoffs which will occur if the corresponding strategies are used.

This is probably the dullest-looking game in the world. At least, it's a fair one. The players have no interesting control over the results; no matter what they do, the payoff is zero. Or is this true?

There is a saying to the effect that murder will out, but it contains no hint as to the speed of retribution: In describing the formation of a game matrix, in Chapter 1, we eased past a point which you are entitled to know when considering this 2 × 2 game.

The matrix is indexed against Blue and Red strategies, i.e., against complete courses of action among which Blue and Red may choose freely. The list is exhaustive, so any accessible sequence of actions, inspired or stupid, that a player can make is represented by one of the strategies. Does it follow that only one number will appear in each box of the matrix? How about chance events—moves by Nature, if you like? It is clear that the use of a specific strategy by Blue and of one by Red is no guarantee that the payoff will be *unique*. For example, the game might be such that the strategies of the players will only determine whether the payoff goes to Blue or to Red and that the magnitude of the payoff is determined by Nature, who spins a roulette wheel to find out how much will be paid.

So what does it mean when the number 0 appears in a box? Or 6, or − 2, or anything else? Well, it may be a very solid value, i.e., the one and only possible outcome when the corresponding pair of strategies is used. But if the game is such that, with fixed and chosen strategies for each player, the payoffs can be any one of several numbers, then 'the payoff,' listed in the diagram, is an average value whose elements de-

pend on chance. This is really important, and we must look at an example to get this notion clear.

Suppose that one pair of strategies really leads to three possible outcomes; in one Blue *wins* 8, in another Blue *wins* 24, and in still another Blue *loses* 8. It will not do to put into the corresponding box of our game matrix a number (8), which is the simple average of the three numbers 8, 24, and −8, because Nature may not be equally fond of them. For example, the chance mechanism might be the toss of a coin; the +8 (gain for Blue) could correspond to heads, the −8 (loss for Blue) to tails, and the +24 (gain for Blue) could correspond to the coin standing on edge. With coins as thin as they are these days, we would neglect the on-edge case (+24), and assume that the others are equally likely; in which case the proper average would be the result of adding +8 and

−8 and dividing by 2 as follows: $\dfrac{8-8}{2} = 0$. This would be the value to use in the matrix. But if the coin used is approximately one-third as thick as it is wide, so that heads, tails, and on-edge will be equally

likely, then it is reasonable to use $\dfrac{8+24-8}{3} = 8$ as the value of the payoff. In general, the proper technique is to *weight the numbers according to the odds which favor their appearances.* For example, if the odds favoring the 8, 24, and −8 were 1:3:4, then the appropriate average would be

$$\frac{1 \times 8 + 3 \times 24 + 4 \times (-8)}{1 + 3 + 4} = \frac{8 + 72 - 32}{8} = 6$$

The average so found is called by mathematicians *the expected value.* It is evident that this is a use of language which requires special care in interpretation. We do *not* expect the value (in this case 6) to turn up when Blue and Red use the strategies which lead to this box—indeed, the payoff '6' is actually impossible of occurrence in this box—but we do expect the average effect to tend toward 6. The import of the expected value may be appreciated by thinking of it in this way: If Blue wants Red to play this box, 6 is a fair side-payment for Blue to make in advance of each play. If any other amount is paid, by either player, the game will be unfair to one player, who will then go broke more often than he should.

Slightly chastened by this necessary side excursion, we return to

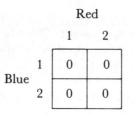

where we now note that there may be chance events—with transitory gains and losses—hidden behind the zeros. We use this to illustrate a weakness (or a hidden requirement) of Game Theory: Suppose the northeast-southwest boxes contain no chance events, so that the actual payoffs on each individual play are exactly what they appear to be, zero. Then suppose that the 0 in the northwest box is really an average payoff, based on two equally likely chance events, worth 1 and −1, respectively, and that the 0 in the southeast box also is based on two chance events, but here worth 1000 and −1000. Since a player's assets are necessarily limited, he may very properly fear Strategy 2 (which might ruin him in one blow), and therefore have a strong preference for Strategy 1, which covers the northwest box. So our use of expected values (i.e., of long-term average values), to compress the description of the chance-event effect, involves the tacit assumption that the player is able and willing to weather temporary vagaries of chance.

Where do we stand if this assumption violates the facts? It isn't likely that the Game Theorist will be helpful: he will claim that the difficulty is ours, not his; that we are in trouble because we were careless about the worth of the alternative payoffs, which is not his province. Since we court ruin as casually as trivial losses, we have failed to use the proper payoff, namely, the real value to the player. The units used in calculating the payoff may have been dollars or lives; in any event, some conventional unit that tricked us into counting instead of facing the hugely more difficult problem of assessing and comparing in nonnumerical terms. To remedy this, one should devise a value scale which better reflects the utility and disutility of the various possible outcomes to the player.

If we do so, however, it is most unlikely that the payoffs to Blue and to Red will continue to be equal in magnitude and opposite in sign—the identifying characteristic of a zero-sum game. So we infer this: If the players prefer one payoff to another of equal average value, the scales need to be corrected; and once they are corrected, it is unlikely that the game will comprise simply an exchange of assets. In other words, it will probably become non-zero-sum; and, as mentioned earlier, we do not know much about such games.

You may feel at this juncture that the Theory of Games is a pretty weak and fragmentary theory. The 'expected' value of a payoff may not always furnish a sensible basis for decision and action; and this fact has forced us, in turn, to realize that we don't have any very good systems for comparing the desirability or undesirability of certain outcomes. All this, moreover, arises before we even get our foot in the door of the theory—before we begin to talk about simple little zero-sum, two-strategy games.

If you are just too discouraged by this, you can, after all, throw down a book even easier than you can turn off a television set. But those with the courage to go on will find that, despite these handicaps, the theory does have some interesting and useful things to say.

We shall assume henceforth that the player has sufficient assets to be able to stand sampling fluctuations, so that expected values can reasonably serve as his guides.

Now consider this game:

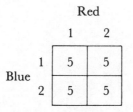

If the strategies used are Blue 1 vs. Red 2, Red must pay Blue 5 units; and the situation is the same for every pair of strategies. So it differs very little from the last game. The important difference is that this is not a fair game for Red; it would become fair if, in addition to the terms of the game itself, Blue agreed to pay Red 5 units each time they play. In this game, as in the previous one, it obviously doesn't matter which strategy is used; for precisely the same payoff occurs no matter what choice of strategies is made. Also it has the property that (military) intelligence is valueless; i.e., foreknowledge of the enemy's plan is not necessary or useful. This, as we shall see, will be a property of many games when good strategies are used by the players.

We note, in passing, that matrixes of the form we have been discussing are similar to those associated with games of chance. In a pure game of chance, the payoff elements not only all look alike, they are in fact identical. This means that if the payoffs are expected values, they are all based on the same numbers and chance events; all of this is in contradistinction to games of skill, or to games of strategy, in which the players have at least some control over the outcome.

To avoid the burden of a possibly unfamiliar object, we shall for a time proceed without negative numbers in the game matrixes. Such games are basically unfair to Red, who just pays and pays. You may think of these games as situations which Red does not voluntarily enter, but which, once in, he wishes to pursue as economically as possible; or as situations in which Red can demand a suitable side-payment, once he knows how to calculate it.

SADDLE-POINTS

We now come to a game in which the players have preferred strategies:

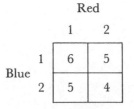

We resort to the fundamental argument of Game Theory: Blue wants to win as much as possible, but dares not be dependent on the largess of Red. He therefore examines each of his strategies in turn, to see just how much he can win even though Red, in effect, may be peeking over his shoulder. He assumes that whatever strategy he chooses, Red will make the best countermove available. Hence Blue lists the worst pay-offs that can result from each of his choices, as follows:

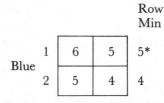

Of these, he prefers the larger—the 5, which we mark by an asterisk. So Blue has some reason to believe that Blue 1 is a good strategy.

Red, on the other hand, examines the columns to see how it may be for him. His basic outlook, including respect for the enemy, is similar to Blue's, except that he wants to keep the payoffs small. As he considers each strategy, he allows for the fact that Blue may discover his interest in it and make the best counterchoice open to Blue. Therefore Red lists the worst payoffs (the largest ones) which may result from his using each strategy:

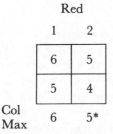

Of these, he naturally prefers the smaller—the 5, marked by an asterisk. So if he uses the strategy Red 2, his losses will never exceed 5.

This is again a coincidence of the kind encountered in the example The Campers: Blue and Red have discovered single strategies which guarantee that the payoff will be some unique number—5, in this case —against an inspired adversary; and each knows that the payoff will be more favorable (than 5) to him if the enemy is not inspired. This is the situation called *saddle-point*.

Generally, *when the larger of the row minima is equal to the smaller of the column maxima, the game is said to have a saddle-point;* and the players should stick to the strategies which intersect at the saddle-point.

To discover that there is a saddle-point, each player must examine the game both from his own and the enemy's point of view. He lists the minimum of each row, and marks (*) the greatest of these; then the maximum of each column, and marks (*) the least of these. If the two marked numbers are equal, the game has a saddle-point, and the players should elect the strategies which correspond to the marked row and column.

MIXED STRATEGIES

Let us try this magic, which led us to a solution in the last example, on another game. Consider this one:

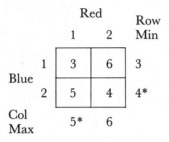

The greatest value for a row minimum is 4, corresponding to the strategy Blue 2; so if Blue adopts this, he will never receive a payoff smaller than 4. The smallest column maximum is 5, for Red 1; so use of Red 1 ensures Red against losses greater than 5.

So there is no saddle-point and we have an unexplored situation. Blue can guarantee himself a payoff of 4 units. Red is sure it won't cost *him* more than 5. You could say that these are the quantities that fairly good players could be sure to produce—only elementary common sense is needed. The range between the quantities is the hunting ground where the master player can pick up something more.

Let us look at Blue's situation again. If he adopts Blue 1 as his strategy, he should reckon on Red's discovering it and driving his income down to 3 units per play; and if he adopts Blue 2, his income may sink to 4 units. However, judging by a similar inspection of Red's situation, Blue should be able to win between 4 and 5 per play, on the average, instead of 3 or 4. Yet the play of Blue 1 or of Blue 2 is the only course of action open to him.

The dilemma itself provides the clue to the solution. Since Blue must do something, and since the steadfast election of either strategy will permit Red to profit unduly, Blue must consider the remaining alternative: to use both strategies. Now in any one play of the game he is constrained to use one or the other strategy, because, by definition, each strategy is a *complete* course of action; and the use of one excludes the other. *So Blue must sometimes use one strategy and sometimes the other.*

In other words, he needs a super or grand strategy which contains the original strategies as elements. In the terminology of Game Theory, a grand strategy is called a mixed strategy, and the element, which we have been calling simply a 'strategy,' is called a pure strategy. A pure strategy is one of the numbered strategies which you stick to for a play of the game: the grand strategy governs your choice of pure strategies.

You will probably notice immediately that this course of action may place you in a situation where security measures are necessary. For in the present game, if Red knows which strategy Blue will really use, he will be in a position to clobber Blue. It will turn out that the *necessary* security limits are clearly marked: Blue must keep secret his decision regarding the strategy he will use in each particular future play of the game. However, *he can permit Red to gain complete knowledge of his past actions, as well as to know his grand strategy.* If it is a good grand strategy, Red cannot prevent Blue from winning all that the game affords for Blue. Similar remarks apply to Red's situation.

We have assumed that the enemy is intelligent and well served; yet it is vital that he not know which pure strategy you will use on the next play. There is one known method which is a sure defense against such an opponent—sure within the limits imposed by the nature of the game itself: namely, to *let the decision regarding strategy depend entirely on some suitable chance event;* so a chance mechanism of some sort is an essential part of a good grand strategy. There will then be no possibility of his gaining useful advance information, for your ultimate course of action will be just as obscure to you as it is to him.

Let's return to the example for a brief demonstration:

		Red	
		1	2
Blue	1	3	6
	2	5	4

Suppose, for the moment, that Red's grand strategy is to make his decisions by tossing a coin—playing Red 1 when the coin shows a head and Red 2 when it shows a tail. When this grand strategy is used against Blue 1, Red will have to pay 3 about half the time and 6 half the time; so the 'expected' or long-term average value of the payoff will be

$$\frac{1 \times 3 + 1 \times 6}{1 + 1} = 4\frac{1}{2}$$

(Here the 1's represent the odds provided by the coin, namely, 1:1. We have weighted each payoff by the odds favoring its occurrence, added together these products, and divided by the sum of the odds.)

Now try the coin-tossing grand strategy against Blue's other alternative, Blue 2: the average payoff is

$$\frac{1 \times 5 + 1 \times 4}{1 + 1} = 4\frac{1}{2}$$

as before. So by leaving the decision to the toss of a coin, Red can make the payoff average 4½ against either Blue strategy. He can therefore average 4½ against any Blue strategy, which is better (remember, Red wants to keep the payoff small) than the 5 he would surely pay by just playing Red 1.

This is an important concept in Game Theory, that of *mixed strategies:* the concept that a player should sometimes use one pure strategy, sometimes another, and that *the decision on each particular play should be governed by a suitable chance device.* We can anticipate that this will be a feature of most games, that it will fail to appear only when circumstances are somewhat special.

You may feel, momentarily, that it is somewhat irresponsible to select a course of action—possibly when the issue is an important one—by use of a chance mechanism. Actually, there is nothing irresponsible about it: all the cogent reasoning which you feel should go into the decision does go into it. It is injected when the problem is formulated, when the payoffs are assessed, and when the odds are computed which govern the chance device and hence the choice of strategy. The chance device is thus an instrument of your will and not your master. The fact that the final instrumentality in the decision process is a machine which does not think deep thoughts is not significant. A bomb isn't very intelligent either; for that matter, the bombardier may on occasion give more thought to blondes than to target selection; of course, as we follow the chain back, it is comforting to suppose that pertinent intellectual activity occurs somewhere.

THE ODDMENT

When we learn, a little later, how to compute mixed strategies, we shall always express the results in terms of *odds*. Thus it may be that Blue should mix Blue 1 and Blue 2 according to the odds 8:5, meaning that in the long run he uses Blue 1 eight times to every five uses of Blue 2. However, we shall frequently be staggering under the burden of a What's-it, unless we have a christening: Henceforth we shall refer to any one number which is at the moment a fugitive from a set of odds as an *oddment*. Thus if the odds are 8:5 and we want to talk about the 8 alone, we shall say that the oddment is 8; Noah Webster convinces us that this usage is almost defensible, legally. Mathematically, an oddment is a monstrous invention, for the odds 8:5 are equivalent to odds of 16:10 (or to 24:15, or to 4:2½, or to infinitely many other pairs), which suggests, for example, that oddment 8 and oddment 16 may be equivalent; so no matter what positive value you assign to an oddment, no one can say it is incorrect. However, whenever we use the word, it is understood that we intend, eventually, to produce a complete set of them, using a comparable scale throughout. The subject comes up of course because in calculating odds we produce one number at a time, and sometimes wish to refer to it.

As a matter of convenience, we shall use zero oddments to indicate that certain strategies are not to be used. Thus 3:0:2 means that a player's first and third strategies are to be mixed according to the odds 3:2, whereas his second strategy is not to be used.

RULES FOR FINDING ODDS

We are not going to give you a *logical* discussion of how to find good mixed strategies, because we don't know how to do it in nonmathematical terms. However, we shall present you with a collection of rules of thumb, which will enable you to compute good mixtures by simple arithmetical steps—simple steps, but sometimes tedious ones. The procedure for 2 × 2 games is this:

Step 1. Look for a saddle-point, which is easy to do. (Just compare the max of the row mins and the min of the column maxes, to see if they are equal.) If there is one, your work is done and the best grand strategy is a pure one, namely, the Blue and Red strategies which intersect at the saddle-point.

Step 2. If there is no saddle-point, the best grand strategies are mixed strategies. Abandon your work on the saddle-point and do the following instead:

We begin by attending to Red's strategy. Consider just Red 1 and 2 and the payoff matrix (we use numbers from the last example):

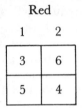

Subtract the numbers in the second row from those in the first, and put the answers in two new boxes

Then the oddment for Red 1 is in the box shaded here:

while the oddment for Red 2 is in the other box, i.e., the one shaded here:

One of these numbers will be negative, always; just neglect the minus sign. The oddments then represent the odds according to which Red should mix his strategies, i.e., 2:2, which indicates an even mixture and clearly is equivalent to 1:1; in other words, Red should mix his strategies by using an equal-odds device, such as a coin.

Note particularly the curious symmetry of the shaded boxes; the oddment for Red 1 is *not* in the Red 1 column.

The above rule of thumb is a little more complicated than it need be, but it sets the stage for some rules we shall need later in larger games; it seems wise, pedagogically, to have you do some of your suffering early. The rule for finding Blue's best mixture is the same, with everything turned sideways. We illustrate it by the same example:

Blue 1	3	6
Blue 2	5	4

Subtract the second column from the first, which yields

Blue 1	−3
Blue 2	1

Then the number in this shaded box,

Blue 1	
	1

and the one in this,

Blue 2	−3

are the oddments for Blue 1 and Blue 2, respectively. Blue should use a mixed strategy based on 1 part of Blue 1 to 3 parts of Blue 2; i.e., he should use the odds 1:3.

We observe in passing that a toss of a coin won't produce these odds —the ones that Blue needs. We shall turn to the question of practical chance devices, capable of giving any desired odds, at the end of this chapter.

The method is so easy there is some danger of beating it to death, but let's do one more example for practice:

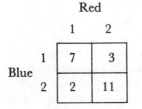

Red

		1	2
	1	7	3
Blue	2	2	11

(This is not a game of craps, despite the numbers.) First, check for a saddle-point:

Red

		1	2	Row Min
	1	7	3	3*
Blue	2	2	11	2
Col Max		7*	11	

There isn't one, as is clear from the fact that the marked (*) numbers are unequal; but it is important that we look for one, however, for if there is one, the job is done. Moreover, *the method we use for finding mixed strategies will usually give false results if applied to a game having a saddle-point.*

So we abandon the work done and begin again, this time by looking for a mixed strategy for Red:

Red

	1	2
	7	3
	2	11

Subtracting the elements of the second row from those of the first, we obtain

Red

	1	2
	5	−8

Therefore the oddment Red 1 is in this shaded box,

Red

1

and that for Red 2 is in this,

Red

2

So the odds are 8:5, favoring Red 1 over Red 2; i.e., he should mix his two strategies, 1 and 2, in the ratio of 8:5. Similarly, we find that Blue should play Blue 1 and 2 according to the odds 9:4.

VALUE OF THE GAME

We have dropped hints, from time to time, that there is some particular quantity of payoff in each game that good play will win for you, against good play by your opponent. This quantity is called the *value of the game*. You cannot on the average win more than the value of the game unless your opponent plays poorly. If it is a positive quantity, then it is the quantity that Blue would have to pay Red at each play in order to make the game fair. Most parlor games, being fair, have the value zero.

In a game with a saddle-point, the value of the game is the same as the value at the saddle. Thus, in

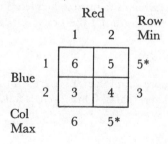

the value of the game is 5. For if the players play properly, i.e., use the strategies Blue 1 vs. Red 2, then the payoff is steadily 5 to Blue.

In 2 × 2 games requiring a mixed strategy, the value of the game is the average payoff which results from use of the best mixture of one player against *either* pure strategy of the other. Thus in

Red

		1	2
	1	3	6
Blue	2	5	4

which we found requires that Red mix his strategies in equal amounts, as by using a coin, his average against Blue 1 is

$$\frac{1 \times 3 + 1 \times 6}{1 + 1} = 4\frac{1}{2}$$

This is the value of the game. If we had made the calculation using Blue's best mixture (recall it was 1:3) against, say, Red 1, the result would be the same, i.e.,

$$\frac{1 \times 3 + 3 \times 5}{1 + 3} = 4\frac{1}{2}$$

In 2 × 2 games which require mixed strategies, the average payoff is always the same (4½ in this game) when the good mixture of one player is tested against either pure strategy of the other. This average payoff is *the value of the game*. We may turn the situation around and use the constancy of the value as a test of accuracy on the mixture: If the presumed good mixed strategies don't yield the same average payoffs when used against each of the enemy's pure strategies, then the presumption is wrong and the mixtures are not the best.

SCALE EFFECTS

Let us try a few more simple games, partly for practice and partly to uncover several general facts. Three games follow:

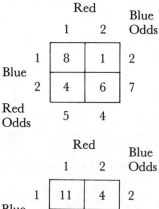

Red
	1	2	Blue Odds
Blue 1	8	1	2
Blue 2	4	6	7
Red Odds	5	4	

Red
	1	2	Blue Odds
Blue 1	11	4	2
Blue 2	7	9	7
Red Odds	5	4	

Red
	1	2	Blue Odds
Blue 1	16	2	2
Blue 2	8	12	7
Red Odds	5	4	

The three games given above, by some strange trick of fate, all call for the same mixtures of strategies: Blue mixes according to the odds 2:7 and Red according to the odds 5:4. Why is it that these should be played alike? Observe that the second may be obtained from the first by adding 3 to each payoff:

8	1
4	6

$+ 3 =$

11	4
7	9

and that the third may be obtained from the first by multiplying each payoff by 2:

$$\begin{array}{|c|c|} \hline 8 & 1 \\ \hline 4 & 6 \\ \hline \end{array} \times 2 = \begin{array}{|c|c|} \hline 16 & 2 \\ \hline 8 & 12 \\ \hline \end{array}$$

These games illustrate a general fact: *The play of a game is not affected by adding a constant to all payoffs or by multiplying all payoffs by a positive constant.*

The values of the game in the three instances are:

$$\frac{5 \times 8 + 4 \times 1}{5 + 4} = 4\%$$

$$\frac{5 \times 11 + 4 \times 4}{5 + 4} = 7\% = 4\% + 3$$

$$\frac{5 \times 16 + 4 \times 2}{5 + 4} = 9\% = 4\% \times 2$$

So, while the *play* is unaffected, *the value of a game is affected when a constant is added to, or multiplied into, the payoffs.*

In physical terms, adding a constant to the payoffs just affects the unfairness level of the game; multiplying the payoffs by a positive constant is equivalent to a change in currency.

GOOD PLAY VS. POOR

We now know just about everything about 2×2 games except this: What happens if one player (say, Blue) plays an optimum strategy and the other (Red) does not? We distinguish two cases, for the moment:

Case 1. If there is a saddle-point (i.e., if the good grand strategies are pure strategies), then Red will act with profligate generosity if he fails to use his optimum pure strategy.

Case 2. If the good grand strategies are mixed strategies, then it doesn't matter what Red does—Blue's good play will keep the average payoff on an even keel. Similarly, if Red plays his proper mixed strategy, the outcome will be the same regardless of how Blue plays. The answer is different—and somewhat more satisfying—for larger games, such as 2 × 3 games.

Up to this point we have manipulated numbers and boxes, just conforming to a set of ground rules. These nonsparkling activities have been referred to—hopefully, you may feel—as games. For a change of pace, we now introduce several examples, which are somewhat less staid. If you have been amazed at the appellation 'game' for the recent activities, you may still be capable of surprise at its use in these examples. The philosophy obviously is (and we hope you will adopt it) that most anything may turn out to be just a game.

EXAMPLE 2. THE HIDDEN OBJECT

Let us dress up the 2 × 2 game in a hypothetical military version: Suppose that a pair of Blue bombers is on a mission; one carries the bomb and the other carries equipment for radar jamming, bomb-damage assessment, or what-have-you. These bombers fly in such a way that Bomber 1 derives considerably more protection from the guns of Bomber 2 than Bomber 2 derives from those of Bomber 1. There is some concern lest isolated attacks by one-pass Red fighters shoot down the bomb carrier, and the survival of the bomb carrier transcends in importance all other considerations. The problem is: Should Bomber 1 or Bomber 2 be the bomb carrier, and which bomber should a Red fighter attack?

The possible strategies are

> Blue 1 = bomb carrier in less-favored position
> Blue 2 = bomb carrier in favored position
> Red 1 = attack on less-favored position
> Red 2 = attack on favored position

Suppose the chance that the bomb carrier will survive, if attacked, is 60 per cent in the less-favored position and 80 per cent in the favored position and is 100 per cent if it is not attacked. Then the situation is this:

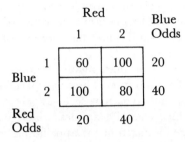

	Red		Blue
	1	2	Odds
Blue 1	60	100	20
Blue 2	100	80	40
Red Odds	20	40	

If we analyze this game just as we have previous ones (and the reader had better check this), it turns out that both Blue and Red should mix their strategies; that both should use the same mixture; and that this mixture should consist of 20 parts of Strategy 1 to 40 parts of Strategy 2. In other words, Blue has a 40-to-20 preference for putting his bomber in the protected position and Red has a 40-to-20 preference for attacking that position. (These odds, 40:20, are of course equivalent to 2:1.) The value of the game to Blue is

$$\frac{20 \times 60 + 40 \times 100}{20 + 40} = 86\frac{2}{3} \text{ per cent}$$

So if Blue uses an appropriate chance device for selecting the bomb-carrier position, instead of doing what comes naturally, he will tend to get 86⅔ per cent of his bombs past the fighters, instead of 80 per cent. The improvement is about 8 per cent.

EXAMPLE 3. THE DAIQUIRIS

In the last example the benefit from a mixed strategy amounted to 8 per cent. You must not get the notion that the gains need be so modest. Consider the following episode, drawn from the archives of a good bar. Alex and Olaf are killing time between flights.

"I know a good game," says Alex. "We point fingers at each other; either one finger or two fingers. If we match with one finger, you buy me a Daiquiri. If we match with two fingers, you buy me two Daiquiris. If we don't match I let you off with the payment of a dime. It'll help to pass the time."

Olaf appears quite unmoved. "That sounds like a very dull game —at least in its early stages." His eyes glaze on the ceiling for a mo-

ment and his lips flutter briefly; he returns to the conversation with: "Now if you'd care to pay me 42 cents before each game, as partial compensation for all those 55-cent drinks I'll have to buy you, then I'd be happy to pass the time with you."

"Forty-one cents," says Alex.

"All right," sighs Olaf. "You really should pay 42 cents, at least once in every 30 games, but I suppose it won't last that long."

In this game the payoff matrix looks like this (identify the strategy Alex 1 with one finger, etc.):

Olaf

		1	2
Alex	1	55	10
	2	10	110

Olaf luckily noticed that this game is unfair to him; so he insisted on a side-payment which would make it fair, provided he played properly.

How should it be played? Let us make the usual calculation:

		Olaf 1	Olaf 2	Alex Odds
Alex	1	55	10	100
	2	10	110	45
Olaf Odds		100	45	

So Olaf should play a mixed strategy, mixing one finger and two fingers according to the odds of 100:45. If he does this, while Alex sticks to one finger, say, his losses per play will average

$$\frac{100 \times 55 + 45 \times 10}{100 + 45} = 41\tfrac{1}{29}$$

Similarly, if Alex sticks to two fingers, Olaf's losses will average

$$\frac{100 \times 10 + 45 \times 110}{100 + 45} = 41\tfrac{1}{29}$$

In fact his losses will average $41\tfrac{1}{29}$ no matter what Alex does; so by using the 100-to-45 strategy-mix (or 20:9), Olaf can stabilize the game, and determine the side-payment he must demand of Alex, to make it a fair game.

The best grand strategy for Alex is also a 100:45 mix, which will establish his average winnings at $41\tfrac{1}{29}$ per play. Note that Alex can only guarantee himself a win of 10 cents by adopting a pure strategy, which Olaf will catch on to. This 10 cents is to be compared with the average of $41\tfrac{1}{29}$ cents the mixed strategy guarantees him. So here we have a case where a mixed strategy is several hundred per cent better than a pure strategy.

In general, if the payoffs in one pair of diagonal boxes are small, and if those in the other pair are large, it is very important that a mixed strategy be used.

EXAMPLE 4. THE RIVER TALE

Steve is approached by a stranger who suggests they match coins. Steve says that it's too hot for violent exercise. The stranger says, "Well then, let's just lie here and speak the words 'heads' or 'tails'—and to make it interesting I'll give you $30 when I call 'tails' and you call 'heads,' and $10 when it's the other way around. And—just to make it fair—you give me $20 when we match."

Warned by the environment (they are on a Mississippi packet), Steve suspects he should have the man arrested, rather than play with him. This question piques his interest more than the game, so he takes the trouble to do this calculation:

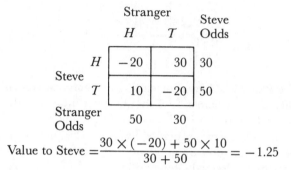

Value to Steve $= \dfrac{30 \times (-20) + 50 \times 10}{30 + 50} = -1.25$

So the best he can hope for is an average loss of $1.25 per play (to guarantee even that, he must call 'heads' and 'tails' in the ratio of 3:5); so his suspicions are confirmed.

EXAMPLE 5. THE ATTACK-DEFENSE GAME

Blue has two installations. He is capable of successfully defending either of them, but not both; and Red is capable of attacking either, but not both. Further, one of the installations is three times as valuable as the other. What strategies should they adopt?

Take the value of the lesser installation to be 1. Then if both survive, the payoff is 4; if the greater survives, the payoff is 3; if the lesser survives, the payoff is 1. Designating the defense (or attack) of the lesser installation as Strategy 1, we have

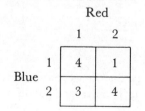

There is no saddle-point, as you may promptly check by writing down the row minima and the column maxima, so a mixed strategy is required. We compute it as usual:

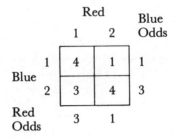

| | | Red | | Blue |
		1	2	Odds
Blue	1	4	1	1
	2	3	4	3
Red Odds		3	1	

Therefore Blue should favor defending his more valuable installation with odds of 3:1; whereas Red should favor attacking the lesser installation with odds of 3:1. The value of the game is 3¼.

This example seems, superficially, very like the bomber-fighter example, but note how different the strategy is; here it is better for Red to attack the less valuable position.

EXAMPLE 6. THE MUSIC HALL PROBLEM

Sam and Gena agree to meet outside the Music Hall at about 6 o'clock on a winter day. If he arrives early and she is late, he will have to drive around the block, fighting traffic and slush, until she appears. He assigns to this prospect a net worth of −1. If she arrives early and he is late, she will get very cold and wet. He estimates his joy-factor in this case as −3. So the game is

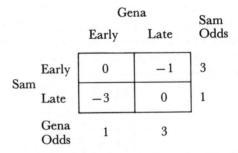

| | | Gena | | Sam |
		Early	Late	Odds
Sam	Early	0	−1	3
	Late	−3	0	1
Gena Odds		1	3	

This game requires a mixed strategy. Sam should play *early* and *late* according to the odds 3:1. The average payoff is −¾.

It is of only academic interest to mention the Game Theory solution of Gena, which is 1:3 with the long odds on *late*. For she of course isn't interested in minimizing Sam's joy; the payoff values are Sam's, and she plays the role of Nature.

EXAMPLE 7. THE DARKROOM

Goldy is in his darkroom developing an irreplaceable negative—the new baby at the age of 8 minutes. He hasn't had much sleep lately and his technique, normally impeccable, is ragged. He hums as he waits for 15 minutes to pass.

He stops humming.

Is this the full-strength developer? Or did he dilute it 2:1 the other day, for prints? He rises to the occasion, first with Anglo-Saxon remarks, and then with analysis.

If the developer is what it should be, 15 minutes will be just right; score 10. If it has been diluted, he can rescue it by relying on the reciprocity law—which is always stated, discouragingly, as Reciprocity Law Failure. If he develops 30 minutes, the quality should fall off only a little, say to 9.

But suppose he develops 15 minutes in a dilute solution? A thin flat negative of a monotone object in a shadowless room—it isn't riches, but something could be done with it; it may be worth 6. Finally, the question of 30-minute development at full strength: grain-size up, contrast like a blueprint, mitigated by a grey pall of chemical fog; with this overage film he's lucky if it's worth 2.

The game matrix is

		Nature		Goldy
		1	2	Odds
Goldy	1	10	6	7
	2	2	9	4

where the strategies are

Goldy 1 = develop 15 minutes
Goldy 2 = develop 30 minutes
Nature 1 = make it full strength
Nature 2 = dilute it

So it appears that a mixed strategy is called for, with odds of 7:4, favoring the shorter period. With it his expectations are

$$\frac{7 \times 6 + 4 \times 9}{11} = 7\frac{1}{11}$$

the value of the game. Strictly, he should play the odds. But if he believes the rise and fall of picture quality are roughly proportional to time, he may prefer to buy some of each and develop for

$$\frac{7 \times 15 + 4 \times 30}{11} = 20\tfrac{5}{11} \text{ minutes*}$$

EXAMPLE 8. THE BIRTHDAY

Frank is hurrying home late, after a particularly grueling day, when it pops into his mind that today is Kitty's birthday! Or is it? Everything is closed except the florist's.

If it is not her birthday and he brings no gift, the situation will be neutral, i.e., payoff 0. If it is not and he comes in bursting with roses,

and obviously confused, he may be subjected to the Martini test, but he will emerge in a position of strong one-upness—which is worth 1. If it is her birthday and he has, clearly, remembered it, that is worth somewhat more, say, 1.5. If he has forgotten it, he is down like a stone, say, −10.

* The implications of this alternative action are discussed later; see page 103.

So he mentally forms the payoff matrix,

		Nature		Frank
		Not Birthday	Birthday	Odds
	Empty handed	0	−10	0.5
Frank	Flowers	1	1.5	10

and sees that the odds are 10:0.5, or 20:1, in favor of flowers.

This example contains an unplanned object lesson, for which the reader must thank a sadist among the early readers of the manuscript. It is unfortunately the case that the writer forgot to test this matrix for a saddle-point, so of course it has one: Frank should always bring flowers, and the reader should always do as we say, not as we do.

The chance that a matrix chosen at random will contain a saddle-point is very substantial for small matrixes. The odds are 2:1 in favor in a 2×2 game, 3:7 in a 3×3, and 4:31 in a 4×4. The chances dwindle to below one in a thousand in a 9×9 game.

EXAMPLE 9. THE HUCKSTER

Merrill has a concession at the Yankee Stadium for the sale of sunglasses and umbrellas. The business places quite a strain on him, the weather being what it is.

He has observed that he can sell about 500 umbrellas when it rains, and about 100 when it shines; and in the latter case he also can dispose of 1000 sunglasses. Umbrellas cost him 50 cents and sell for $1; glasses cost 20 cents and sell for 50 cents. He is willing to invest $250 in the project. Everything that isn't sold is a total loss (the children play with them).

He assembles the facts regarding profits in a table:

Selling when it

		Rains	Shines	Odds
Buying for	Rain	250	−150	5
	Shine	−150	350	4

and immediately takes heart; for this is a mixed-strategy game, and he should be able to find a stabilizing strategy which will save him from the vagaries of the weather.

Solving the game, he finds that he should buy for rain or for shine according to the odds 5:4, and that the value of the game is

$$\frac{5 \times (250) + 4 \times (-150)}{9} = \$72.22$$

Rather than play the odds,* he decides to invest five-ninths of his capital in rainy-day goods and four-ninths in sunny-day goods. So he buys $161.11 worth of umbrellas (including $22.22 from the sunny-day program) and $88.89 worth of sunglasses, and he prepares to enjoy the steady profit of $72.22.

EXAMPLE 10. THE SQUAD CAR

This is a somewhat more harrowing example. The dispatcher was conveying information and opinion, as fast as she could speak, to Patrol Car 2, cruising on the U.S. Highway: "... in a Cadillac just left Hitch's Tavern on the old Country Road. Direction of flight unknown. Suspect Plesset is seriously wounded but may have an even chance if he finds a good doctor, like Dr. Haydon, soon—even Veterinary Paxson might save him, but his chances would be halved. He shot Officer Flood, who has a large family."

Deputy Henderson finally untangled the microphone from the riot gun and his size 14 shoes. He replied:

---

* The implications of this alternative action are discussed later; see page 103.

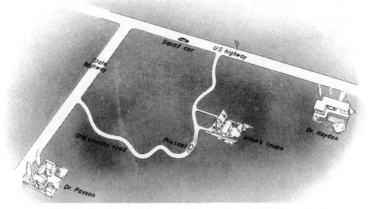

"Roger. We can cut him off if he heads for Haydon's, and we have a fifty-fifty chance of cutting him off at the State Highway if he heads for the vet's. We must cut him off because we can't chase him —Deputy Root got this thing stuck in reverse a while ago, and our cruising has been a disgrace to the Department ever since."

The headquarters carrier-wave again hummed in the speaker, but the dispatcher's musical voice was now replaced by the grating tones of Sheriff Lipp.

"If you know anything else, don't tell it. He has a hi-fi radio in that Cad. Get him."

Root suddenly was seized by an idea and stopped struggling with the gearshift.

"Henderson, we may not need a gun tonight, but we need a pencil: this is just a two-by-two game. The dispatcher gave us all the dope we need."

"You gonna use *her* estimates?"

"You got better ones? She's got intuition; besides, that's information from headquarters. Now, let's see . . . Suppose we head for Haydon's. And suppose Plesset does too; then we rack up one good bandit, if you don't trip on that gun again. But if he heads for Paxson, the chances are three out of four that old doc will kill him."

"I don't get it."

"Well, it don't come easy. Remember, Haydon would have an even chance—one-half—of saving him. He'd have half as good a chance with Paxson; and half of one-half is one-quarter. So the chance he dies must be three-quarters—subtracting from one, you know."

"Yeah, it's obvious."

"Huh. Now if we head for Paxson's it's tougher to figure. First of all, *he* may go to Haydon's, in which case we have to rely on the doc to kill him, of which the chance is only one-half."

"You ought to subtract that from one."

"I did. Now suppose he too heads for Paxson's. Either of two things can happen. One is, we catch him, and the chance is one-half. The other is, we don't catch him—and again the chance is one-half—but there is a three-fourths chance that the doc will have a lethal touch. So the over-all probability that he will get by us but not by the doc is one-half times three-fourths, or three-eighths. Add to that the one-half chance that he doesn't get by us, and we have seven-eighths."

"I don't like this stuff. He's probably getting away while we're doodling."

"Relax. He has to figure it out too, doesn't he? And he's in worse shape than we are. Now let's see what we have."

Cad goes to

		Haydon	Paxson
Patrol car goes to	Haydon	1	¾
	Paxson	½	⅞

"Fractions aren't so good in this light," Root continued. "Let's multiply everything by eight, to clean it up. I hear it doesn't hurt anything."

Cad

		Haydon	Paxson
Patrol car	Haydon	8	6
	Paxson	4	7

"It is now clear that this is a very messy business . . ."

"I know."

"There is no single strategy which we can safely adopt. I shall therefore compute the best mixed strategy."

	Cad Haydon	Paxson	Patrol car Odds
Patrol car Haydon	8	6	3
Paxson	4	7	2
Cad Odds	1	4	

"Umm . . . three to two in favor of Haydon's. Read the second hand on your watch. Now!"

"Twenty-eight seconds."

"Okay. To Haydon's. This thing would be a lot easier to steer backwards if it didn't have such a big tail . . ."

"That magic with the second hand is sure silly," groused Deputy Henderson. "Why didn't we just head for Haydon's in the first place?"

"Well, it would have affected our chances—in this case not much, but some. If we were just simple-minded about Haydon's, Plesset could sucker us by going to Paxson's and have a 75 per cent risk. This way, no matter how he figures it, he runs an 80 per cent risk. About the watch: the first 30 seconds and the next 20 seconds go with the 3-to-2 odds."

SUMMARY OF 2 × 2 METHODS

We shall summarize the technical material of the first part of this chapter by working one more example.

Consider the game

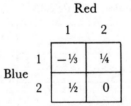

Suppose we wish to know *how to play it*. Suppose also that we are allergic to fractions and to negative numbers. We discovered that the *play* of a game is unaffected when the numbers in the matrix are multiplied by a constant, or when a constant is added to each element. To alleviate the allergy then, let us multiply the above by a felicitously chosen constant, say 12, which improves its appearance no end:

Red

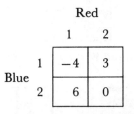

		1	2
	1	−4	3
Blue	2	6	0

Now we get rid of the negative number by adding something to each element. The number 4 will do, giving us

Red

		1	2
	1	0	7
Blue	2	10	4

We are now ready to seek the solution. We of course begin by looking for a saddle-point:

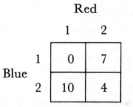

		Red 1	Red 2	Row Min
	1	0	7	0
Blue	2	10	4	4*
Col Max		10	7*	

The maxmin is 4 and the minmax is 7. There is no saddle-point; so we abandon the work and begin again. Starting with an investigation of Red's problem,

Red

	1	2
	0	7
	10	4

we subtract each element from the one above, which gives us

Red

so the oddments for Red 1 and Red 2 are in these boxes:

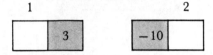

One is negative, but we always disregard the sign of an oddment; there-fore Red's good grand strategy (if nothing has gone wrong) is 3:10.

Blue's odds are the next problem:

Blue
	1	0	7
	2	10	4

Subtracting each element from the one on its left, we have

Blue
	1	−7
	2	6

so the oddments are in these boxes:

1
	6

2
−7	

Blue's mixture is according to the odds 6:7—if all is well.

Let us see if all is well by trying these mixtures against each of the enemy's pure strategies. Blue's 6:7 grand strategy, when used against Red 1, yields

$$\frac{6 \times 0 + 7 \times 10}{6 + 7} = \frac{70}{13}$$

and against Red 2 it yields

$$\frac{6 \times 7 + 7 \times 4}{6 + 7} = \frac{70}{13}$$

On the other hand, Red's 3:10 mix yields, against Blue 1,

$$\frac{3 \times 0 + 7 \times 10}{3 + 10} = \frac{70}{13}$$

and, against Blue 2,

$$\frac{3 \times 10 + 10 \times 4}{3 + 10} = \frac{70}{13}$$

So everything is fine. Blue and Red should play these mixed strategies, 6:7 and 3:10, respectively.

Is this $^{70}/_{13}$ the value of the game? No! Recall that we multiplied by 12 and then added 4 to each element. We can get the value by the process of just unwinding the steps: Since we *added* 4, we now subtract 4; i.e., $\frac{70}{13} - 4 = \frac{70 - 52}{13} = \frac{18}{13}$. Then, since we *multiplied* by 12, we now divide by 12; i.e., $\frac{18}{13} \div 12 = \frac{18}{13 \times 12} = \frac{3}{26}$. So the value of the game is $^3/_{26}$, which may be verified by trying our mixed strategies in the original game. Take, say, the Blue 6:7 mix against Red 2, which yields

$$\frac{6 \times \frac{1}{4} + 7 \times 0}{6 + 7} = \frac{3}{26}$$

as it should.

We tuck in at this juncture a set of Exercises, for those who like to exercise—it is probably the best way to fix new ideas. The answers are in the back of the book somewhere.

EXERCISES 1

Determine oddments and values of the following games:

1. Red

		1	2
Blue	1	7	4
	2	1	2

2. Red

		1	2
Blue	1	5	8
	2	6	7

3.

Red

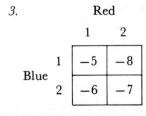

		1	2
	1	−5	−8
Blue	2	−6	−7

4.

Red

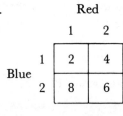

		1	2
	1	2	4
Blue	2	8	6

5.

Red

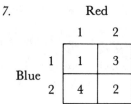

		1	2
	1	1	2
Blue	2	4	3

6.

Red

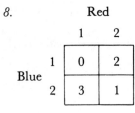

		1	2
	1	1	2
Blue	2	3	4

7.

Red

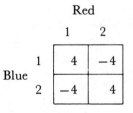

		1	2
	1	1	3
Blue	2	4	2

8.

Red

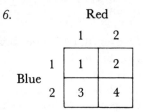

		1	2
	1	0	2
Blue	2	3	1

9.

Red

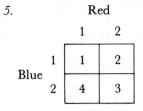

		1	2
	1	−1	1
Blue	2	2	0

10.

Red

		1	2
	1	−3	1
Blue	2	3	−1

11.

Red

		1	2
	1	4	−4
Blue	2	−4	4

12.

Red

		1	2
	1	4	−5
Blue	2	−5	4

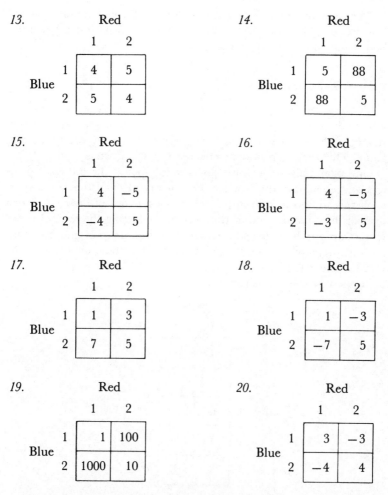

13. Red

		1	2
Blue	1	4	5
	2	5	4

14. Red

		1	2
Blue	1	5	88
	2	88	5

15. Red

		1	2
Blue	1	4	−5
	2	−4	5

16. Red

		1	2
Blue	1	4	−5
	2	−3	5

17. Red

		1	2
Blue	1	1	3
	2	7	5

18. Red

		1	2
Blue	1	1	−3
	2	−7	5

19. Red

		1	2
Blue	1	1	100
	2	1000	10

20. Red

		1	2
Blue	1	3	−3
	2	−4	4

PART TWO: *2 × m Games*

Games in which one player has *two* strategies and the other has *many* are cunningly called 'two-by-*m* games'; this is written as '2 × *m* games'—the letter *m* may represent any whole number greater than

2. The remainder of this chapter will be devoted to such games. Fortunately, their solution requires practically nothing beyond a knowledge of how to solve 2 × 2 games.

SADDLE-POINTS

These games always have either a saddle-point solution (i.e., a good *pure* strategy) or a good mixed strategy based on *two* pure strategies. So, in effect, the 2 × *m* game may be reduced to a 2 × 2 game. This reduction may be effected in various ways.

One should always look first for a saddle-point; the process is painless and concludes the work if there is one. Recall that we inspect each row to find its minimum, select the greatest of these, and then inspect the columns to find the maxima, selecting the least of these. If the two numbers (called, incidentally, the maxmin and minmax) are equal, there is a saddle-point, and their common value is the value of the game.

Suppose we have this game:

	Red 1	Red 2	Row Min
Blue 1	4	4	4
Blue 2	5	3	3
Blue 3	6	5	5*
Blue 4	1	3	1
Blue 5	5	4	4
Col Max	6	5*	

It has a saddle-point, Blue 3 vs. Red 2, because the maxmin and minmax are equal (to 5).

DOMINANCE

If there is no saddle-point, then examine the strategies of the player who has many strategies. It may be self-evident that some of them are so inferior that they should never be used. Suppose that Blue is the player with many: If one of his strategies is superior to another, *on a box-by-box basis,* then the former is *dominant,* and the latter should be eliminated from the matrix. In fact, this elimination may be made even when the 'superior' one is in part equal to the other. As an example, consider

Red

	1	2
1	2	5
2	4	3
Blue 3	3	6
4	5	4
5	4	4

On comparing just Blue 1 and Blue 3, i.e.,

	1	2	5
Blue			
3	3	6	

it is clear that Blue 3 is dominant. Similarly, comparing Blue 2 and 4, and Blue 4 and 5, i.e.,

	2	4	3			4	5	4
Blue					Blue			
	4	5	4			5	4	4

we see that Blue 4 is dominant in both instances; so the game may be reduced, for calculation, to

Red

	1	2
3	3	6
Blue 4	5	4

Similarly, if Red has many strategies, and if one column dominates another on a box-by-box basis, then the dominant column may be eliminated (recall, Red wants small payoffs).

The utility of this principle in simplifying games is obvious. Of course it may be used in a 2 × 2 game; but in that case if there is a dominant row or column, there is also a saddle-point.

MIXED STRATEGIES

If there is no saddle-point, examine the strategies of the player who has many for dominance. If the player with more than two strategies is Blue, eliminate the *dominated* ones; if the many-strategy player is Red, drop the *dominant* ones. The game which remains (which of course may still be just the original game, for there may be no dominance) then really contains a 2 × 2 game which has this property: Its solution is also a solution to the 2 × *m* game. The *theory* of how to find that critical 2 × 2 game is painfully simple, although the *practice* may be very tedious: just look for it.

That means, take one of the 2 × 2's, solve it, and try the solution in the original game. If the mix for the 2-strategy player holds up against each of the strategies of the many-strategy player, you've discovered the solution. 'Holds up' means that the 2-strategy player does at least as well (usually better) against any of the other fellow's strategies as he does against the pair that appear in the 2 × 2 subgame.

Let's try to point up the last remark a little better. In the 2 × 2 games which called for mixed strategies, we found that a player using a good mixture would win the same quantity against either of his opponent's two strategies. We said it wasn't a very satisfying state of affairs, but that something better was coming. Now it has. In the 2 × *m* game, one still wins the same quantity against either of the strategies which appear in the opponent's best mix, but one wins more against the opponent's remaining strategies; there are exceptions, where the winnings are the same against some or all the remaining strategies, too, but usually one gets more.

Perhaps some examples will help. Take

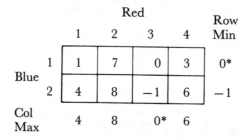

		Red				Row Min
		1	2	3	4	
Blue	1	1	7	0	3	0*
	2	4	8	−1	6	−1
Col Max		4	8	0*	6	

Here the maxmin and the minmax are equal (to zero), and so the game has a saddle-point. It's a fair game, because its value is zero. The best strategies are Blue 1 vs. Red 3.

Let's take another, say,

		Red							Row Min
		1	2	3	4	5	6	7	
Blue	1	−6	−1	1	4	7	4	3	−6
	2	7	−2	6	3	−2	−5	7	−5*
Col Max		7	−1*	6	4	7	4	7	

Here the maxmin is −5 and the minmax is −1; so there is no saddle-point. However, Red 3, 4, 5, and 7 dominate Red 2; so the game may be reduced to

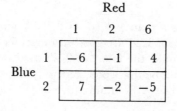

		Red		
		1	2	6
Blue	1	−6	−1	4
	2	7	−2	−5

in which we must search for a 2 × 2 game whose solution satisfies the 2 × 3 game (and which, incidentally, will then automatically satisfy the original 2 × 7 game).

We start with the first one,

Red

		1	2
	1	−6	−1
Blue	2	7	−2

which, by methods tried and true, we find has the solution 9:5 for Blue and 1:13 for Red. The value of this game is

$$\frac{9 \times (-6) + 5 \times 7}{9 + 5} = -\frac{19}{14}$$

when Blue plays 9:5 against Red 1. It is of course the same when used against Red 2; i.e.,

$$\frac{9 \times (-1) + 5 \times (-2)}{14} = -\frac{19}{14}$$

The test of the pudding will come when we try the Blue 9:5 mix against the remaining Red strategy (Red 6)—the yield must be greater than (or equal to) $-\frac{19}{14}$. One should always be hopeful at this point; if it is not a solution, the disappointment will be with us long enough, after we discover it. Against Red 6, this Blue mix will win

$$\frac{9 \times 4 + 5 \times (-5)}{9 + 5} = \frac{11}{14}$$

which is certainly larger than $-\frac{19}{14}$; so the 9:5 mix for Blue and the 1:13:0:0:0:0:0 mix for Red are good strategies for them to use in this game. In this example, the first 2 × 2 we tried worked. You can plainly see that life won't be that good, generally. If we had not eliminated several strategies by the dominance argument, there would have been $\frac{7 \times 6}{2}$ 2 × 2's in this game, and we might have had to try all 21 of them before meeting with success. (In a 2 × m game, there are $\frac{m(m-1)}{2}$ games of the 2 × 2 variety.)

GRAPHICAL SOLUTION

While we don't want to burden you with special tricks, we can't resist giving one here, just to make the 2 × m livable. And make the most of this, for you won't have things like this later, in larger games, when you will need them more! Using the last example, i.e.,

Red

		1	2	3	4	5	6	7
	1	−6	−1	1	4	7	4	3
Blue	2	7	−2	6	3	−2	−5	7

plot the payoffs (−6 and 7) of the first Red strategy on separate vertical axes and connect the points, as shown at the left below. Do the same for all of Red's strategies, all on the same graph, as shown at the right:

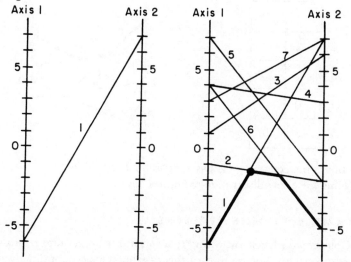

Quite a mess, of course, but strictly utilitarian. Now make double weight the line segments which bound the figure from *below;* then find and mark with a dot the *highest point* on this double-weight boundary. The lines which intersect at the dot identify the strategies Red should use in his mixture.

In the last graph, sections of lines (1) and (2) fulfill these conditions, so that Red's good mixture is based on Red 1 and 2, which agrees with our earlier finding. Once the significant pair is identified, it is easy to compute, from the 2 × 2, the good strategies for Blue and Red in the 2 × m game. Note that this single graph has permitted us to identify instantly (after an hour of searching for the ruler) the significant one among twenty-one 2 × 2 games.

If you are dealing with an m × 2 game, in which *Blue* has many strategies, the graphical method for identifying the significant pair of Blue strategies is similar to the above: Mark the line segments which bound the graph from *above,* and dot the *lowest point* on this boundary; the lines which pass through the dot identify the critical strategies for Blue.

For example, the graph of this game,

Red

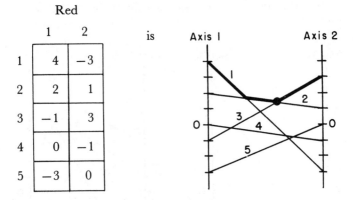

	1	2
1	4	−3
2	2	1
3	−1	3
4	0	−1
5	−3	0

is

So Blue's mixed strategy is based on Blue 2 and Blue 3.

Time for a few illustrative examples again.

EXAMPLE 11. THE SPELLERS

Goldsen and Kershaw, owing to a natural bent, get into a battle of words. Kershaw, hoping to turn this to economic account, finally suggests that they create words, or try to, according to the scheme he outlines:

"Suppose you choose either the letter a or the letter i, and, independently, I will choose f, t, or x. If the two letters chosen form a word, I will pay you \$1, plus a \$3 bonus if the word is a noun or pronoun. In the rare event that the letters chosen don't form a word, you pay me \$2."

"I never go into a thing like that," says Goldsen, "without a deep analysis of the what-where-when aspects, including the game-theoretic. Pardon me a moment."

He writes out the payoff matrix,

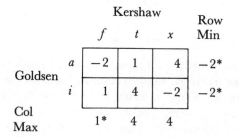

and examines it for a saddle-point; but the greatest row minimum (-2) does not equal the least column maximum (1); so the matrix has no saddle-point. Before looking for a mixed strategy, he notices that Kershaw's strategy f is dominated by his strategy t, i.e., for Kershaw, t is worse than f, box by box, and therefore it will be conservative to assume that Kershaw won't use t. This leaves for consideration the matrix

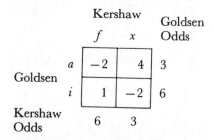

which readily yields the information that Goldsen should choose a and i, at random, in the mixture of 3 to 6 (or 1 to 2). Kershaw, who proposed the game, knows that he should choose f, t, or x according to the odds 6:0:3; i.e., he should choose twice as many f's as x's, and no t's.

Goldsen still hasn't decided he wants to play this game.

"Wait till I compute the value, old chap," he mumbles. "I may find a parameter, or something. Now, let's see: if he uses f, I'll average about $\dfrac{1 \times (-2) + 2 \times 1}{3} = 0$—not much, but not fatal. If he uses t,

I'll get $\dfrac{1 \times 1 + 2 \times 4}{3} = \$3$; and I hope he does. If he uses x, I'll get

$\dfrac{1 \times 4 + 2 \times (-2)}{3} = 0$. Umm . . . this looks like an eminently fair

game. I'm surprised at you. Shall we play?"

EXAMPLE 12. THE SPORTS KIT

Two Muscovite guards—call them A and B, though those are not their names—obtain at the canteen a sports kit. This contains a carton of Pall Malls, a revolver, a bullet, and rules for Russian Roulette. (The most interesting thing we've uncovered in working up this example is that Russians often call it 'French Roulette,' that Frenchmen call it 'Spanish Roulette,' and so on;* but the Russians would

doubtless claim the invention if they appreciated the benefits that would follow its adoption as the official Party sport.)

Each player antes a pack of cigarettes. As senior officer present, A plays first. He may add two packs to the pot and pass the gun to B; or he may add one pack, spin the cylinder, test fire at his own head, and (God willing) hand the gun to B. If and when B gets the

---

* A cosmopolitan reader informs us the phenomenon is quite general; e.g., what are called wieners in Frankfurt are known as frankfurters in Vienna.

gun, he has the same options: he may hand it back, adding two packs, of course; or he may add one pack and try the gun. The game is now ended. Each picks up half the pot, if that is feasible; otherwise the bereaved picks it all up.

Player A has two strategies: (1) he may pass or (2) he may gamble. Player B has four strategies: (1) he may pass, ignoring A's decision; or (2) he may gamble, ignoring it; or (3) he may mimic A; or (4) he may do the opposite of what A does.

The calculation of the payoff matrix is elementary, fussy, and boring to read and to write; therefore we leave the verification as an exercise. (This is ingenious!)* It turns out to be

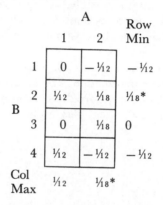

		A 1	A 2	Row Min
B	1	0	$-\frac{1}{12}$	$-\frac{1}{12}$
	2	$\frac{1}{12}$	$\frac{1}{18}$	$\frac{1}{18}$*
	3	0	$\frac{1}{18}$	0
	4	$\frac{1}{12}$	$-\frac{1}{12}$	$-\frac{1}{12}$
Col Max		$\frac{1}{12}$	$\frac{1}{18}$*	

The maxmin and minmax are equal; so the game has a saddle-point—at B2–A2. This means that A should always gamble and B should always gamble. And the game is unfair to A.

---

* You may feel that it is also dishonest, for we promised that we would confine the book to elementary arithmetic, and this is probably not the first time that you have detected something stronger—such as probability theory—lurking about. Actually, the book is based, fundamentally, on quite a lot of mathematics. Even those which we have called the 'crank-turning' aspects have been discussed, heretofore, principally in very technical journals, in articles which you might not recognize as communication among English-speaking persons. And the process of abstraction and modelmaking, by which we go from the real world to the game matrix, may require a deep knowledge of the subject matter of the particular problem, much mathematical versatility, some ingenuity, and occasional late hours.

We try to meet these two problems as follows: The Game Theory aspects are discussed and strictly arithmetical procedures are set forth. The physical problem is discussed and the game matrix is deduced by simple argument, when possible; otherwise, we use (secretly) whatever tools we need to use, and present you with the finished product, the matrix.

This game is somewhat more than just another 2 × *m*, for it has this special property: the players have complete information about past events at the times they make their decisions. Analysis has in fact been carried out by mathematicians, which proves that *every game of complete information has a saddle-point;* so it isn't surprising that this one has. They may or may not also have good mixed-strategy solutions.

Chess is another game with complete information. Hence there is a way to play it—a pure strategy—which is at least as good as any mixed strategy. We don't know what this saddle-point strategy is, nor whether it leads to a win for White or for Black, or to a draw—the game is too large and complicated for analysis. Direct enumeration of the strategies is quite impossible, since the estimated number (of strategies) contains more zeros than you could write in a lifetime. If we did know the solution, there would be no point in playing the game.

Bridge and Poker, which are also beyond present-day analysis, are not games of complete information. The outcome of chance events—the deal of the cards—is only partially known to the participants during the play of the hand. Good play undoubtedly requires a mixed strategy.

EXAMPLE 13. THE HI-FI

The firm of Gunning & Kappler manufactures an amplifier having remarkable fidelity in the range above 10,000 cycles—it is exciting comment among dog whistlers in the carriage trade. Its performance depends critically on the characteristics of one small, inaccessible condenser. This normally costs Gunning & Kappler $1, but they are set back a total of $10, on the average, if the original condenser is defective.

There are some alternatives open to them: It is possible for them to buy a superior-quality condenser, at $6, which is fully guaranteed; the manufacturer will make good the condenser and the costs incurred in getting the amplifier to operate. There is available also a condenser covered by an insurance policy which states, in effect, "If it is our fault, we will bear the costs and you get your money back." This item costs $10.

Their problem reduces to this 3 × 2 game:

		Nature	
		Defect	No Defect
	(1) Cheap	− 10	− 1
Gunning & Kappler	(2) Guarantee	− 6	− 6
	(3) Insure	0	− 10

By graphical examination (or by trial, if you prefer),

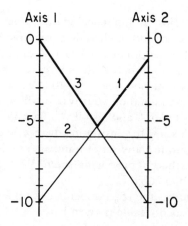

we find that (1) and (3) are the pertinent strategies; so we calculate the odds for them:

		Nature		G & K
		Defect	No Defect	Odds
G & K	1	− 10	− 1	10
	3	0	− 10	9

The odds are 10:9 for this 2 × 2; so their mixed strategy for the original game is 10:0:9. That is, each time that they install a condenser they should leave it to chance whether they use a cheap $1 condenser or one of the expensive insured ones, with the odds weighted slightly in favor of the former. They should not buy the $6 condenser.

The value of the game is

$$\frac{10 \times (-10) + 9 \times 0}{10 + 9} = -5\frac{5}{19}.$$

Thus the average amount they may have to spend per amplifier is approximately $5.26.

CHANCE DEVICES

You may wonder, just as a practical matter, where to find a chance device that will produce odds such as those required by mixed-strategy games. A coin will serve for odds of 1 to 1. A die will produce 1 to 5, 2 to 4, and so on up to 5 to 1. A deck of cards can produce things like 1 to 12 (e.g., Ace vs. non-Ace), 3 to 10 (e.g., face card vs. nonface card), etc. The second hand on your watch will serve very well for a random number between 0 and 59—if you haven't been peeking at it lately, so you won't know the answer in advance, even approximately. But a game may require any set of numbers as odds; so you should have a general method of producing them. We shall therefore do a brief excursion on this subject.

A device that meets the requirements is a table of random numbers. A sample of such a table is given here:

$$
\begin{array}{ccccc}
35 & 07 & 53 & 39 & 49 \\
56 & 62 & 33 & 44 & 42 \\
36 & 40 & 98 & 32 & 32 \\
57 & 62 & 05 & 26 & 06 \\
07 & 39 & 93 & 74 & 08 \\
68 & 98 & 00 & 53 & 39 \\
14 & 45 & 40 & 45 & 04 \\
07 & 48 & 18 & 38 & 28 \\
27 & 49 & 99 & 87 & 48 \\
35 & 90 & 29 & 13 & 86 \\
\end{array}
$$

The fact that these digits are set off in groups of two has no signifi-cance other than antieyestrain. You can start at any point in a large table of this kind and read up or down, right or left, taking the digits in groups of 1, 2, 3, or more, always with the same result: nonsense, unexpectedness, randomness. While a lot of thought goes into the production of such numbers, there is supposed to be none embodied in them. (This is part of a table, containing a million random digits, produced by a sort of super electronic roulette wheel. A small section of that table is reproduced in the Appendix.)

It would be unwise to suppose that you can, reliably, write down a random sequence of digits, out of your head. Habits, prejudices, orderliness, and so on, all militate against its being random; the best pedigree you could have, to make the effort a sound one, would be that of a perfect imbecile, in the full medical sense.

How are such tables used to produce the odds we desire? Take the odds 5:2, for illustration: Decide where you will start in the table by stabbing with a finger a few times to get page, row, and column numbers; start there. Perhaps you have the tenth digit in the eighth column of page 1. If that digit is a 0, 1, 2, 3, or 4, play Strategy 1. If it is a 5 or 6, play Strategy 2. If it is a 7, 8, or 9, skip it and take the digit immediately below (or above, but decide in advance) and use it to determine which strategy to employ. For the next play of the game (if there is another), take the digit below the last one used and proceed as before, playing Strategy 1 if the digit is a 0, 1, 2, 3, or 4, playing Strategy 2 if it is a 5 or 6, and moving to a new entry in the table (up or down, as previously decided) if it is a 7, 8, or 9. Similarly, if the odds that concern you are 85 to 9, decide on the location of a *pair* of digits to be used. If they form a number in the range 00 to 84, play Strategy 1; if in the range 85 to 93, play Strategy 2; if they form 94 to 99, forget them and take the next pair.

Minor modifications of this method are sometimes desirable. For instance, odds of 7 to 4, which total 11, require that you use a two-digit random number, which ranges from 00 to 99; but any number between 11 and 99 will be rejected, and so the method wastes a lot of numbers. If you are wealthy in random digits, it still costs you time to scan and reject so many numbers. The solution in this case is: Divide the largest two-digit number, 99, by the sum of the oddments, 11. The answer is exactly 9 in this instance; in general, it will be a whole number plus a fraction. Now multiply each of the oddments by that whole number, 9; this yields $7 \times 9 = 63$ and $4 \times 9 = 36$, which is a good form in which to use the oddments. You would now adopt Strategy 1 when the table shows a number in the range 0 to 62, and Strategy 2 for numbers in the range 63 to 98. You would reject the number 99.

SUMMARY OF 2 × *m* METHODS

Before leaving you with some Exercises, we shall review briefly the technique of solving 2 × *m* games. To this end, consider the game

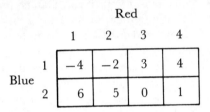

Red

		1	2	3	4
Blue	1	−4	−2	3	4
	2	6	5	0	1

Is there a saddle-point? We look at the maxmin and minmax:

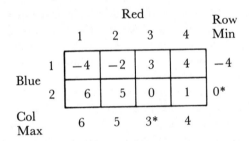

		Red				Row Min
		1	2	3	4	
Blue	1	−4	−2	3	4	−4
	2	6	5	0	1	0*
Col Max		6	5	3*	4	

These are 0 and 3, so there is no saddle-point; we abandon that effort.

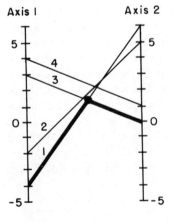

Axis I   **Axis 2**

We note that Red 4 is worse than Red 3, box by box; so it could be eliminated immediately. However, we intend to seek the significant pair of strategies graphically, and the dominated Red 4 will be eliminated automatically. Plotting the four Red strategies, we get the graph at the left.

The highest point on the lower envelope is at the intersection of Red 1 and Red 3; so these are the pertinent strategies. The subgame based on these is

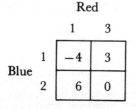

It does not have a saddle-point; so we compute the mixed strategies. The oddments for Blue are based on

and those for Red are based on

Red

	1	3
	−10	3

so Blue's best mixture is 6:7, while Red's is 3:10. Returning to the original 2 × 4 game, we see that Blue should use the grand strategy 6:7, while Red should use 3:0:10:0.

The value of the game is found by playing Blue's good grand strategy, 6:7, against *either of the strategies in Red's good mix,* i.e., against Red 1 or Red 3. Using Red 1, we find the value to be

$$\frac{6 \times (-4) + 7 \times 6}{6 + 7} = \frac{18}{13}$$

EXERCISES 2

Determine the oddments and values of the following games:

1.

Red

		1	2	3
Blue	1	7	0	3
	2	2	−1	−6

2.

Red

		1	2	3	4
Blue	1	−2	−6	−5	−1
	2	−3	−7	−6	−2

3.

Red

		1	2	3
Blue	1	4	0	2
	2	6	7	1

4.

Red

	1	2	3	4	5	6	7	8
Blue 1	11	18	24	30	24	20	21	23
2	24	30	24	23	21	17	18	20

5.

Red

	1	2	3	4
Blue 1	2	2	3	4
Blue 2	4	3	2	2

6.

Red

	1	2	3	4
Blue 1	10	7	11	0
Blue 2	−8	−2	−9	1

7.

Red

	1	2	3	4	5
Blue 1	5	−2	3	7	0
Blue 2	0	5	2	2	6

8.

Red

	1	2	3	4	5
Blue 1	1	−3	5	−7	9
Blue 2	−2	4	−6	8	−10

9.

Red

	1	2	3	4	5
Blue 1	−1	3	−5	7	−9
Blue 2	2	−4	6	−8	10

10.

Red

	1	2	3	4	5
Blue 1	9	−5	7	1	−3
Blue 2	−10	4	−8	−6	2

11.

Red

	1	2	3	4
Blue 1	8	0	6	7
Blue 2	3	6	3	1

12.

Red

	1	2	3
Blue 1	4	6	0
Blue 2	3	0	7

13.

Red

	1	2	3	4	5	6
Blue 1	0	9	8	7	4	2
Blue 2	10	1	2	3	6	8

14.

Red

	1	2
Blue 1	1	3
Blue 2	5	7
Blue 3	9	11

15. Red

	1	2
1	1	5
2	2	4
Blue 3	3	3
4	4	2
5	5	1

16. Red

	1	2
1	−1	−3
Blue 2	−5	−7
3	−9	−11

17. Red

	1	2
1	−2	3
Blue 2	3	−2
3	0	0

18. Red

	1	2
1	−2	3
Blue 2	3	−2
3	1	1

19. Red

	1	2
1	−1	5
2	−3	1
3	0	−3
Blue 4	−3	0
5	1	−3
6	5	−1

20. Red

	1	2
1	9	1
2	7	5
Blue 3	8	3
4	5	9
5	6	7

Three-strategy Games
PART ONE: *3 × 3 Games*

MORALE-BUILDING DISCOURSE

It will become very tedious to explain how to solve large games (10 × 10, for instance), with our feet resting only on arithmetic, rather than on the lofty ground of higher mathematics. The difficulties may compare to those in teaching an intelligent—but completely unlettered—child how to render in Spencerian script the names of the American Indian tribes, the instruction all taking place over a telephone, in a somewhat strange dialect, spoken by another child. Solving the 3 × 3 may compare to teaching, by the same means, how to print CAT—itself no mean feat for either party. It is hoped that author and reader don't become alienated in the course of the project.

Before we begin on 3 × 3's, it may be useful to review briefly the general outline of the problem-solving technique we are developing. Faced with a conflict situation which offers several alternatives of action, how do you apply Game Theory to it?

Well, first, you list *all* options (sequences of choices) that are jointly open to you and the enemy—and you must be sure to include chance events that Nature may tuck in, if the possibility is there. Taken literally, this first step is usually impossible, but it may become possible and be adequate if you include just those sequences of possible actions which appear to be highly significant. It is at this point that the art and judgment of the mathematical modelmaker are most severely tested: any fool can list more factors in a given situation than the gods could analyze—for instance, Newton should really have taken account of the fact that the planets are oblate spheroids, that some have mountains, that at least one has trees, and termites, and other pleasant and unpleasant things—and real discrimination is needed to decide where to break off. In practice, you must always break off when you *judge* that the model has enough factors for you to learn something about your problem from it.

After listing the strategies of the players, you must calculate, or estimate, or guess the payoffs which are associated with each pair—one Blue and one Red strategy. Sometimes these payoffs are specifically stated as part of the rules of the game. Sometimes you must compute them by means of probability theory or some other scheme of logical deduction. Sometimes (and this happens embarrassingly often in practical cases) you must just use your best judgment in estimating these payoffs, or in guessing them. But you must have them, and have them in specific numerical form, usually, before you can get forward. (The actual latitude in these matters is discussed in several examples throughout the book, and in some detail in the two sections following page 193.)

When the payoffs are known, the game matrix may be written and you are ready to begin the analysis. What is the purpose of the analysis? To find out how best to play the game, and to assess its value, i.e., to estimate how much you can expect to win or lose if you and the enemy play it well. 'How to play' is synonymous with 'which strategy to use,' or with 'which strategies to choose among, using a suitable random device (whose characteristics you must determine) to identify the chosen.'

And now, back to the crank. Some 3 × 3 games are easy to solve and some are not. You are condemned to work through a battery of methods every time you meet a new one, sustained only by the sure knowledge that there *is* a solution.* Fortunately, much of it is the same as 2 × 2 work. Unfortunately, the part that is not the same will have to be learned by rote, as is common in primers. This is regrettable, but unavoidable, we believe, for quite a lot of mathematics must be concealed by simple procedures. After all, when you learned how to find the square root of a number, some years ago, you probably did not have a glimmer of an idea as to why the procedure you were taught actually worked—at least the present writer didn't understand it. And few operators of motor vehicles have deep knowledge of internal-combustion engines.

---

*After sampling this chapter, the reader may wish to turn to Chapter 6 for a general method of solving games.

SADDLE-POINTS

The first thing to do is look for a saddle-point, and hope there is one. The process is short and painless. For example, the game may look like this:

		Red		
		1	2	3
	1	3	0	2
Blue	2	−4	−1	3
	3	2	−2	−1

Run through and pick out the minimum in each row, and mark the greatest one. Similarly, pick out the maximum in each column, and mark the least. This gives you:

		Red			Row Min
		1	2	3	
	1	3	0	2	0*
Blue	2	−4	−1	3	−4
	3	2	−2	−1	−2
Col Max		3	0*	3	

If these two marked numbers are the same, as they are here, your work is done, for they identify a good strategy for each player. When there is a saddle-point, each player should use the pure strategy which corresponds to it—Blue 1 and Red 2, in the present case. Moreover, the marked numbers are equal to the value of the game. This happens to be zero here, which indicates it is a fair game for all concerned. Note that neither player can be hurt so long as he sticks to his best strategy, but that he will be hurt if he alone departs from the path of virtue.

DOMINANCE

Now let us consider this game:

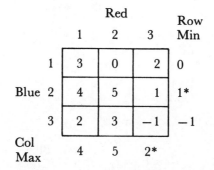

		Red 1	Red 2	Red 3	Row Min
Blue	1	3	0	2	0
	2	4	5	1	1*
	3	2	3	−1	−1
Col Max		4	5	2*	

Just a collection of numbers, which looks neither better nor worse than the last one, so we try to find a saddle-point. The rows yield a maxmin of 1 and the columns a minmax of 2; so we have nothing for our pains. Well, not quite nothing, for *these two numbers tell us that the value of the game is greater than* 1 *and less than* 2.

Let us examine it for dominant strategies. That is, is there any pure strategy which it is patently stupid for Blue (or Red) to use?

Blue wants the payoff to be large. Note that Blue 2 is better, box by box, than Blue 3. So why consider Blue 3, ever? Let's cut it out. Similarly, Red, who wants small payoffs (to Blue), sees that Red 1 is always worse than Red 3; so we drop Red 1.

This leaves us with a 2 × 2 game, namely,

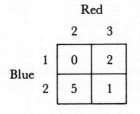

		Red 2	Red 3
Blue	1	0	2
	2	5	1

which we know how to solve; i.e.,

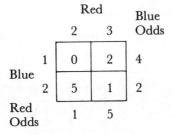

So the solution to the 3 × 3 game is:

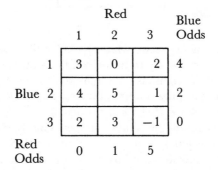

Blue should mix his strategies according to the odds 4:2:0, and Red should mix his according to the odds 0:1:5. The oddment 0 means that the strategy is not employed, of course.

If we had found only one strategy to eliminate by dominance, we should have been left with a 2 × 3 game to solve, instead of a 2 × 2; but we have learned how to solve 2 × 3's, too.

As you see, the concept of Dominance, just as that of Saddle-point, is the same in 3 × 3 games as it was in 2 × m's. The implications are these: If the 3 × 3 has a saddle-point, a good grand strategy is a good pure strategy. If there is no saddle-point, but there is a dominant row or column (or both), then a good grand strategy is a mixed strategy based on two pure strategies.

VALUE OF THE GAME

The average value of a 3 × 3 game—the value which good play will guarantee you in the long run—is also found by a natural extension of the 2 × 2 method: Calculate the average payoff when, say,

Blue's good *mix* is used against one of the *pure* strategies in Red's *best mix*. Thus, in the last example, Blue will average (against Red 2)

$$\frac{4 \times 0 + 2 \times 5 + 0 \times 3}{4 + 2 + 0} = \frac{5}{3}$$

It is the same against Red 3. Moreover, Red's good mix holds his losses to ⅗ against Blue 1, and against Blue 2.

If either player uses his one forbidden strategy—Blue 3 and Red 1 —the results deteriorate. Thus a transgression by Red yields Blue

$$\frac{4 \times 3 + 2 \times 4 + 0 \times 2}{4 + 2 + 0} = \frac{10}{3}$$

and if Blue slips, the payoff becomes

$$\frac{0 \times 2 + 1 \times 3 + 5 \times (-1)}{0 + 1 + 5} = -\frac{1}{3}$$

i.e., Red manages to win ⅓ in a game which is basically unfair for him. You can see that we are working up to games where a poor player cannot depend on his opponent's good play to keep the game in balance for him, as was the case for 2 × 2's with mixed strategies. In more complicated games, the fact that your opponent uses his head by no means excuses you for not using yours. On the contrary —and as you would expect—you have to pay a price for being foolish.

THREE ACTIVE STRATEGIES

So far, we have discussed the solution of 3 × 3's having one active strategy (under Saddle-point), and of some having two active strategies (under Dominance). We now divulge a technique for writing the solutions to 3 × 3 games which have three active strategies. Be sure to test for saddle-point and dominance before using the following, *as it won't work if either of them will;* like surgery, this method is a last resort (almost) procedure, to be used only if medicines fail.

Consider this game:

Red

		1	2	3
	1	6	0	6
Blue	2	8	−2	0
	3	4	6	5

We test for saddle-point and dominance. No luck. It's time for the new rule of thumb.

To begin with, let us confine our attention to Red and determine the frequency with which he should play his various strategies. (I know, you are Blue; but we have to do the work for both Blue and Red, and we feel it is just a little easier to start with Red.) First, subtract each row from the preceding row, i.e., each payoff from the payoff directly above it, and write the results in a new set of boxes:

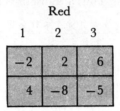

The oddment for Red 1 is found from the numbers in the following shaded boxes (obtained by striking out the Red 1 column in the last matrix):

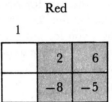

The numerical value of this foursome is the difference between the diagonal products,

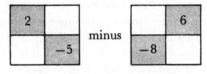

that is,

$$2 \times (-5) - 6 \times (-8) = 38$$

This number, 38, represents the oddment Red 1. If it had turned out to be -38, we should have neglected the minus sign and still claimed 38 as the number sought; but be very careful to keep all signs until you finish computing the oddment.

In exactly the same manner, we may obtain the oddments for Red 2 and Red 3. For Red 2, we strike out the Red 2 column of the shaded matrix, leaving

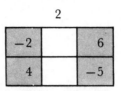

Red

2

from which we see that the oddment Red 2 is governed by the foursome

−2	6
4	−5

Here the difference of the diagonal products is

$$(-2) \times (-5) - 4 \times 6 = -14$$

so the Red 2 oddment is 14. Similarly, we find 8 for Red 3. Collecting results, we have the odds 38:14:8, which describes Red's best mixed strategy.

We get Blue's best mixed strategy in a comparable way, working with the columns instead of rows. Subtracting each column from the preceding one, i.e., each payoff from the payoff on its left, the original matrix yields:

1	6	−6
Blue 2	10	−2
3	−2	1

and the oddment Blue 1 is governed by this:

1		
Blue	10	−2
	−2	1

which has the value $10 \times 1 - (-2) \times (-2) = 6$. Similarly, we find -6 and 48 for Blue 2 and 3; so the best mixed strategy is 6:6:48.

Now we have what we hope * is the solution, namely,

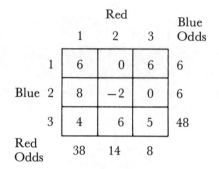

		Red			Blue
		1	2	3	Odds
	1	6	0	6	6
Blue	2	8	−2	0	6
	3	4	6	5	48
Red Odds		38	14	8	

Common factors may be eliminated from both sets of odds. Thus those for Blue contain the common factor 6, and when we divide by it, we get the simpler (but equivalent) set 1:1:8. Similarly, we can suppress a factor 2 in Red's odds, and write them as 19:7:4.

The last step in the three-active-strategies procedure—*and it is an essential step*—is to try out these sets of odds to see whether they have the characteristics of a solution: Recall that a player's best mixture should yield the same payoff against each pure strategy the enemy *should* use, which means all of them in this instance. The calculations are easy.

If Blue plays his 1:1:8 mix against Red 1, his average winnings are

$$\frac{1 \times 6 + 1 \times 8 + 8 \times 4}{1 + 1 + 8} = \frac{23}{5}$$

Against Red 2 also he collects

$$\frac{1 \times 0 + 1 \times (-2) + 8 \times 6}{1 + 1 + 8} = \frac{23}{5}$$

and against Red 3,

$$\frac{1 \times 6 + 1 \times 0 + 8 \times 5}{1 + 1 + 8} = \frac{23}{5}$$

So far, so good. Now try Red's 19:7:4 mix against Blue 1. He loses

---

* The weak word 'hope' is used to cover the possibility of numerical errors and a more fundamental difficulty that is discussed in the next section.

$$\frac{19 \times 6 + 7 \times 0 + 4 \times 6}{19 + 7 + 4} = \frac{23}{5}$$

and the same losses occur against Blue 2 and Blue 3.

We have solved the game, and its value is $2\frac{3}{5}$, or $4\frac{3}{5}$. We know we have solved it, because the average payoff is the same when either player's best mix is played against any pure strategy which appears in the other fellow's best mix.

GAMES WE WISH YOU'D NEVER MET

It turns out that the foregoing battery of methods for solving 3×3 games falls flat on its face, from time to time. Which is to say, there are subtleties in some game payoff-matrixes which affect the generality of the methods offered to you.

As an example and just in passing now:* You will recall some discussion of dominance. Some pure strategies are obviously inferior to others; so they should not be used. Well, sometimes there is a pure strategy which is inferior to a felicitous *mixture* of others, and hence it should not be used. If one of these cases turns up, our methods fail. There are other troubles, similar in nature to this one, which can cause failure. But we can fix them all.

A failure of the 3×3 method given last is always associated with a forbidden strategy: For some reason—probably obscure—one or another of the pure strategies should be abandoned forthwith. And this is the clue to our procedure—pedestrian for sure, but it gets there.

Drop a strategy (any one) and solve the remaining game; then try this solution in the original game. If it works, you have conquered. If it fails, try dropping one of the other strategies. Again solve the remaining game and try the solution in the original. This has to succeed, eventually.

Let's try one:

Red

		1	2	3
	1	6	0	3
Blue	2	8	−2	3
	3	4	6	5

* This subject is expanded in a later section, page 186.

This looks no worse than the last example; in fact, only two of the numbers are changed. We proceed as usual: no saddle-point, no dominance. So we busy ourselves with the three-active-strategies arithmetic—described in the last section—and get, for Blue, 0:0:0; and for Red, 4:4:8. Trouble leaps to the eye this time, for it appears that Blue doesn't want to play any strategy, which is simply not allowable, for we are trying to find out how they should play if they do play. We are lucky that it is obvious that something is wrong, for it could have happened that the odds wore the garments of respectability, such as 3:7:5, but that they were in fact spurious; this would have shown up only when we tried to verify the solutions by calculating the value of the game, etc.

As a trial corrective measure, leave out Blue 1. This gives us the following game:

Red

	1	2	3
2	8	−2	3
3	4	6	5

Blue

We look for a 2 × 2 solution to this 2 × 3 game. Beginning with

Red

	1	2
2	8	−2
3	4	6

Blue

we find one solution of the reduced game, namely,

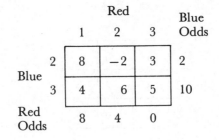

	Red 1	Red 2	Red 3	Blue Odds
Blue 2	8	−2	3	2
Blue 3	4	6	5	10
Red Odds	8	4	0	

We try this in the original game (omitting common factors in the odds):

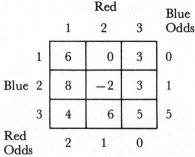

	Red 1	Red 2	Red 3	Blue Odds
Blue 1	6	0	3	0
Blue 2	8	−2	3	1
Blue 3	4	6	5	5
Red Odds	2	1	0	

Is it a solution? Well, Blue wins, against Red 1, the amount

$$\frac{0 \times 6 + 1 \times 8 + 5 \times 4}{0 + 1 + 5} = 4\tfrac{2}{3}$$

and the same against Red 2 and Red 3. (It wouldn't have been surprising if Blue had done better than $4\tfrac{2}{3}$ against Red 3; it just happens he doesn't.) On the other hand, Red, against Blue 1, loses

$$\frac{2 \times 6 + 1 \times 0 + 0 \times 3}{2 + 1} = 4$$

and this is permissible, for Red will usually do better than the value of the game when Blue uses a strategy that does not appear in the good Blue mix. Similarly, we find that Red loses $4\tfrac{2}{3}$ against Blue 2 or Blue 3.

Thus we have a solution. Either player can guarantee, by sticking to the calculated mixture (0:1:5 and 2:1:0, respectively), to keep the average payoff at $4\tfrac{2}{3}$ if the enemy uses one of the pure strategies which occur in his good mix and to do at least as well as $4\tfrac{2}{3}$ if the enemy uses a dubious strategy.

It is very unfortunate that all of this has taken so long to explain, when it's so simple to do. If you are still with us, you have (in terms of the analogy given at the beginning of the chapter) learned to print CAT, in spite of formidable communications.

Perhaps we should review a point which has appeared from time to time: Good play, i.e., use of the proper mixed strategy by Blue (for instance), sometimes provides an insurance policy for both players, since many schemes of play adopted by Red do as well as his best mixed strategy would do. But the insurance policy is ironclad only for Blue, the player using the mixed strategy. Red can really coast against the best mix only when *his* (Red's) best mixture (which we suppose he isn't using) contains *all* the pure strategies; Red may then, with impunity, play any pure strategy against the best Blue mixture, whatever that happens to be. But if certain pure strategies are not allowed in Red's best mixture, then he dare not use one of those *verboten* pure strategies against the good Blue mix.

So use of mixed strategies does not automatically and fully ensure the enemy against needless loss. Rather, it automatically and fully ensures you against needless loss, always. And, in addition, it exacts penalties from the enemy in cases where the enemy has bad strategies available and uses them. It does not, unfortunately, guarantee that these extra penalties will be as heavy as possible.

EXAMPLE 14. SCISSORS-PAPER-STONE

One of the games analyzed early in the history of Game Theory was Scissors-Paper-Stone, a game played by children.* The solution is intuitively obvious; so it was interesting to work toward a known result.

The two players simultaneously name (or characterize through pantomime) one of the three objects. If both name the same object, the game is a draw. Otherwise, superiority is based on the fact that Scissors cuts Paper, that Stone breaks Scissors, and that Paper covers Stone. The payoff matrix is therefore this:

Red

		Scis-sors	Paper	Stone
	Scissors	0	1	−1
Blue	Paper	−1	0	1
	Stone	1	−1	0

* My indefatigable researchers point out a Chinese version: Man eats rooster, rooster eats worm, worm eats man.

There is neither a saddle-point nor a dominant strategy. Red's mixed strategy is found from this matrix:

Red

Scis-
sors Paper Stone

1	1	−2
−2	1	1

obtained by subtracting each row from the preceding one. The oddment for Scissors is based on

Red

Scis-
sors

	1	−2
	1	1

that is,

$$1 \times 1 - 1 \times (-2) = 3$$

For Red's Paper, this diagram

Red

Paper

1		−2
−2		1

yields

1	−2
−2	1

and then

$$1 \times 1 - (-2) \times (-2) = -3$$

For Red's Stone, this diagram

Red

Stone

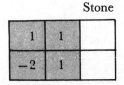

leads to

$$1 \times 1 - 1 \times (-2) = 3$$

By symmetry, the results for Blue will be the same; so the players should play each of the three pure strategies with the same odds, i.e., 1:1:1. (Since the method may fail, such results must always be checked.)

EXAMPLE 15. THE COAL PROBLEM

On a sultry summer afternoon, Hans' wandering mind alights upon the winter coal problem. It takes about 15 tons to heat his house during a normal winter, but he has observed extremes when as little as 10 tons and as much as 20 were used. He also recalls that the price

per ton seems to fluctuate with the weather, being $10, $15, and $20 a ton during mild, normal, and severe winters. He can buy now, however, at $10 a ton.

What to do? Should he buy all, or part, of his supply now? He may move to California in the spring, and he cannot take excess coal with him. He views all long-range weather forecasters, including ground hogs,*dimly.

He considers three pure strategies, namely, to buy 10, 15, or 20 tons now and the rest, if any, later. The costs of the various alternatives are easily found to be

		Mild	Normal	Severe	Row Min
	10	−100	−175	−300	−300
Stockpile	15	−150	−150	−250	−250
	20	−200	−200	−200	−200*
Col Max		−100	−150	−200*	

This game has a saddle-point, corresponding to the 20-ton stockpile.

(If Hans has a little general climatological information, he can use it to advantage. We shall digress on this type of situation later in the chapter. See The Bedside Manner, page 122.)

EXAMPLE 16. THE HEIR

George inherits a fortune and must pay his Uncle Sam $100,000 in inheritance taxes. However, he has one year in which to make the payment. He naturally decides to wait the full year, to get as much revenue as he can through investment. He appeals to Novick and Co., in whose ability he has complete faith, for facts and figures relevant to his problem. It supplies him with these estimates of the gains he

---

*The degree of poetic license permitted the writer by his manuscript critics may be appreciated from this accurate comment: Ground hogs do no weather forecasting on summer afternoons, particularly not on sultry ones.

might make by investing in various ways, each dependent on the political atmosphere, about which Novick and Co. does not care to prognosticate:

In case there is

	War	Cold War	Peace
Bonds	2,900	3,000	3,200
War Babies	18,000	6,000	−2,000
Mercantiles	2,000	7,000	12,000

George somehow recognizes that this is a game between himself and Nature. He is the maximizing player and She—to the extent She is interested—is the minimizing player.

The game has no saddle-point, and no dominance; so we try our usual method for solving 3 × 3's—and get nonsense, i.e., odds that do not have the characteristics required for a solution. So we drop down to the 2 × 2's and, after a little exploration, try

	War	Cold War
War Babies	18,000	6,000
Mercantiles	2,000	7,000

which yields the odds 5:12 for George's strategies (and 1:16 for Nature's). And this solution works in the original game, so we have

	War	Cold War	Peace	George Odds
Bonds	2,900	3,000	3,200	0
War Babies	18,000	6,000	−2,000	5
Mercantiles	2,000	7,000	12,000	12
Nature Odds	1	16	0	

George may play the odds, or he may invest $\frac{5}{17}$ of the $100,000 in War Babies and $\frac{12}{17}$ in Mercantiles. The value of the game is about $6700.

In this example (as well as in some earlier ones) we have done something which appears at variance with our own rules: namely, we have countenanced an interpretation of the odds as a *physical mixture* of strategies on a single play of the game. That is, we have permitted the player to use a little of this strategy and a little of that strategy, instead of insisting that he use just one, basing his choice on a *suitable* chance device.

Indeed, it is evident that we have violated some principle; otherwise the physical mixture would not be possible. Any possible set of actions should be represented by some pure strategy; so the possibility of using a physical mixture should not arise.

This anomalous situation is traceable to the fact that there is an infinite game which is closely related to the finite game stated above. In the infinite game the player could invest, in infinitely many ways, in mixtures of securities—we are ignoring the practical limitations concerning the divisibility of securities and money. Moreover, the partial payoff from each security is proportional to the amount purchased and the total payoff is just the sum of the partial ones. Such situations may be analyzed as infinite games, which turn out to have saddle-points, or as finite games, which turn out to require mixed strategies (usually). By interpreting the latter as a physical mixture, we arrive at a solution which is equivalent to the saddle-point of the associated infinite games.

EXAMPLE 17. THE CATTLE BREEDERS' SEPARATION

Dalkey and Kaplan were going out of the cattle-breeding business, because Kap*lan* felt that he was being edged more and more into the background. They disposed of their stock through normal channels, except for three prize Angus bulls: Ch. Dal*lan*, Ch. Dalk*an* and Ch. Dalk*en*.

These presented something of a problem, for they were valuable beasts—they would easily bring 20, 30, and 40 thousand, respectively —yet the owners could not tolerate the thought of others having

them; nor was either anxious to pay his partner for this large capital gain. Each would count it a gain to own them and a loss if the other had them. It took them some time to work out a mutually agreeable scheme.

They finally decided on this: Each would take, from a deck of cards, a deuce, a trey, and a four—symbolic of bids of 20, 30, and 40 thousand, respectively. One of the bulls would be placed on the block and each of the partners would select and play a card symbolizing his 'bid.' The high bidder would be given free title to the bull; after which, another animal would be considered, bid for by selecting cards from those remaining, and so on. The bulls would be taken in random order. A tie would be resolved by shooting the beast. How should they play their hands? That is, in what order should they play their cards, or should it be left in some way to chance?

The analysis of this is a little messy; we shall just sketch how it goes. It involves any one of three 3 × 3 games, depending on which animal comes up first; and each payoff element in these 3 × 3's in turn is the value of a 2 × 2 game. To see how it goes, suppose old Dallan (worth 2 units) comes up first. And suppose both play their deuces; poor fellow—but neither partner would count it a gain or a loss. If Dalkan came next, then Dalken, the possibilities would be these:

Dalkey plays

3, then 4   4, then 3

	3, then 4	4, then 3
Kaplan plays 3, then 4	0 + 0	−3 + 4
4, then 3	3 − 4	0 + 0

This game, i.e.,

0	1
−1	0

has a saddle-point at the upper left; so its value is zero. (The result is the same if the last two beasts come in the other order.) Then this zero, added to the outcome of the first move, which was also worth zero, is the upper left corner of the Dallan-first 3×3:

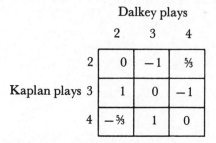

Dalkey plays

	2	3	4
Kaplan plays 2	0	$-\frac{1}{2}$	$2\frac{3}{7}$
Kaplan plays 3	$\frac{1}{2}$	0	$-\frac{1}{2}$
Kaplan plays 4	$-2\frac{3}{7}$	$\frac{1}{2}$	0

The other boxes are filled in similarly. So Kaplan and Dalkey should select their first bids so as to win this game. This requires a 1:23:1 mix, for both players. Then subsequent bids are governed by a 2×2 game. The value of the over-all game is zero; so it's fair.

If Dalkan comes up first, the game turns out to be

Dalkey plays

	2	3	4
Kaplan plays 2	0	-1	$\frac{5}{3}$
Kaplan plays 3	1	0	-1
Kaplan plays 4	$-\frac{5}{3}$	1	0

and if Dalken comes up first,

Dalkey plays

	2	3	4
Kaplan plays 2	0	$-1\frac{3}{5}$	$-\frac{1}{5}$
Kaplan plays 3	$1\frac{3}{5}$	0	$-1\frac{3}{5}$
Kaplan plays 4	$\frac{1}{5}$	$1\frac{3}{5}$	0

requiring strategies of 3:5:3 and 0:0:1, respectively.

The principal value of the example—aside from providing you with a neat scheme for disposition of cattle with which you are emotionally involved—is that it shows how vastly complicated the analysis of a seemingly simple problem may be. Except in situations which are really simple, structurally, expert hands (and probably strong backs) are needed.

EXAMPLE 18. THE DATE

David and John want to date Ann. Neither knows just when she will be home; in fact, the odds are equal that she will arrive at 3, 4, or 5 o'clock. She has a natural preference for Dave; so if John phones first, she will stall for a few minutes to give Dave a chance to catch up; but if Dave doesn't show rather soon, Ann will call John back and accept his invitation. It doesn't promise to be a riotous evening in either case, because each boy has only a dime—which he will squander on the telephone company.

The payoff matrix is built up from elementary probability notions. For instance: If Dave phones at 3 o'clock and John at 5 o'clock, Dave will have a ⅓ chance of getting the date and John will have a ⅔ chance. As a payoff to Dave, this is $\frac{1}{3} - \frac{2}{3} = -\frac{1}{3}$. The complete matrix is:

John calls at

		3	4	5
	3	⅓	0	−⅓
David calls at	4	0	⅔	⅓
	5	⅓	−⅓	1

After fruitlessly checking for saddle-point and dominance, we find that this yields readily to the three-active-strategies method. (The arithmetic is made easy by first multiplying everything by 3.) David should play the odds 2:2:1, whereas John should play 10:4:1. The value of the game to David is ⅕.

This positive value for David results of course from the favoritism shown by Ann. It takes a little digging to discover its meaning in terms of dates for Dave, dates for John, and evenings when Ann defrays only her own expenses. It turns out that the odds favoring each of these events is 97:52:76. So if this game were repeated for 225 nights and the events exactly coincided with the odds, Dave would get 97 dates and John 52; and Ann would not have an escort on 76 nights. Notice that $\dfrac{97 - 52}{225} = \dfrac{1}{5}$; which we hope you find a satisfying check.

SUMMARY OF 3 × 3 METHODS

There are two essential new points of technique used in solving 3 × 3 games which were not required in solving 2 × 2's. These are (a) elimination of strategies by dominance arguments and (b) the shaded-box ritual associated with three active strategies. The general procedure for solving a 3 × 3 game is:

1. Look for a saddle-point. That is, compare the greatest row minimum (for short, the *maxmin*) and the smallest column maximum (the *minmax*); if these are equal, there is a saddle-point.
2. If there is no saddle-point, try to reduce the game by dominance arguments. The numbers in some row may be too small to be interesting to Blue, or those in some column may be too large to be interesting to Red; if so, cut off the offending row or column, and solve the 2 × 3 game (or the 3 × 2 game) which remains.
3. If the two efforts described above fail, give it the shaded-box treatment; i.e., assume that the solution involves all three strategies and try to compute the best mixture, *and test the result*.
4. If the three efforts described above fail, assume that the solution of the 3 × 3 game is the same as the solution of one of the 2 × 2 games contained in it. Try them, one after another, until you succeed or until you have tried them all.
5. If the four efforts described above fail, you have made an error somewhere. Too bad. The Exercises which follow contain opportunities for errors.

Determine the oddments and values of the following games:

1.

Red

		1	2	3
	1	1	4	7
Blue	2	1	7	4
	3	4	7	4

2.

Red

		1	2	3
	1	1	2	3
Blue	2	8	9	4
	3	7	6	5

3.

Red

		1	2	3
	1	1	2	1
Blue	2	2	7	4
	3	1	0	6

4.

Red

		1	2	3
	1	6	−10	3
Blue	2	4	4	4
	3	7	11	−5

5.

Red

		1	2	3
	1	0	1	1
Blue	2	1	0	1
	3	1	1	0

6.

Red

		1	2	3
	1	−1	1	0
Blue	2	0	−1	1
	3	1	0	−1

7.

Red

		1	2	3
	1	0	2	1
Blue	2	2	0	2
	3	1	2	0

8.

Red

		1	2	3
	1	1	1	3
Blue	2	1	3	2
	3	3	2	2

9.

Red

		1	2	3
	1	3	5	2
Blue	2	0	6	8
	3	4	1	3

10.

Red

		1	2	3
	1	7	0	−5
Blue	2	0	1	4
	3	−5	3	6

11.

Red

		1	2	3
	1	0	1	2
Blue	2	1	0	1
	3	2	1	0

12.

Red

		1	2	3
	1	4	3	2
Blue	2	3	4	3
	3	2	3	4

13.

Red

		1	2	3
	1	−6	5	0
Blue	2	2	1	3
	3	1	2	0

14.

Red

		1	2	3
	1	0	4	1
Blue	2	−7	2	1
	3	6	−5	−1

15.

Red

		1	2	3
	1	−7	4	2
Blue	2	0	2	1
	3	6	−5	−1

16.

Red

		1	2	3
	1	1	4	7
Blue	2	1	7	4
	3	5	7	4

17.

Red

		1	2	3
	1	1	3	2
Blue	2	0	7	4
	3	3	8	1

18.

Red

		1	2	3
	1	3	1	6
Blue	2	2	2	0
	3	8	0	3

19.

Red

		1	2	3
	1	1	3	5
Blue	2	2	4	4
	3	6	1	1

20.

Red

		1	2	3
	1	3	1	6
Blue	2	2	4	2
	3	1	5	4

21.

Red

		1	2	3
	1	6	4	3
Blue	2	0	2	8
	3	1	6	3

22.

Red

		1	2	3
	1	8	0	7
Blue	2	3	3	8
	3	3	6	1

23.

Red

		1	2	3
	1	3	0	7
Blue	2	4	6	0
	3	3	4	3

24.

Red

		1	2	3
	1	203	403	103
Blue	2	303	3	103
	3	3	103	303

25.

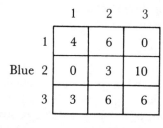

Red

	1	2	3
1	4	6	0
Blue 2	0	3	10
3	3	6	6

26.

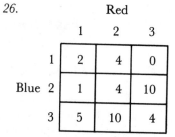

Red

	1	2	3
1	2	4	0
Blue 2	1	4	10
3	5	10	4

27.

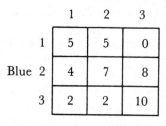

Red

	1	2	3
1	5	5	0
Blue 2	4	7	8
3	2	2	10

28.

Red

	1	2	3
1	3	6	0
Blue 2	5	3	2
3	2	1	6

29.

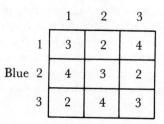

Red

	1	2	3
1	3	5	0
Blue 2	4	4	1
3	1	1	6

30.

Red

	1	2	3
1	3	2	4
Blue 2	0	4	1
3	1	3	4

31.

Red

	1	2	3
1	3	2	4
Blue 2	4	3	2
3	2	4	3

PART TWO: *3 × m Games*

METHOD OF SOLVING

The 3 × m games—in which one player has three pure strategies and the other has many—are solved using the methods given for 3 × 3 games, just as the 2 × m games were solved by 2 × 2 methods. Every 3 × m game has a 3 × 3 solution, a 2 × 2 solution, or a saddle-point. Your problem will be to find the appropriate 3 × 3 (or 2 × 2) subgame, solve it, and demonstrate that the odds thus produced constitute a solution of the original game.

You begin, of course, by looking for a saddle-point—'of course' because it is very easy to do, and because you cannot afford not to know it is present when it is. If there is no saddle-point, then look for dominance, so that obviously unnecessary rows and columns may be stricken from the analysis. If you can eliminate in this way one of the strategies of the player who has but three, the task immediately reduces to a search for a 2 × 2 solution to a 2 × m game.

If the 3 × m game does not have a saddle-point, and if elimination of strategies by dominance has failed to reduce it substantially, then you may have quite a time of it; you must search through all the possible 3 × 3's and 2 × 2's until you find one whose solution is also a solution to the 3 × m game.

Since we burdened ourselves, in the last chapter, with a graphical trick for uncovering the significant 2 × 2 in a 2 × m, we may as well capitalize on it: Search for a 2 × 2 solution to the 3 × m by breaking the latter into 2 × m's. Consider Blue 1 and 2 against all m of Red's strategies; then Blue 2 and 3, against the m; finally Blue 3 and 1.

If there is a 2 × 2 whose solution satisfies the original game, you can find it quite easily by the graphical method. Because it is easy to find, you should probably look for it before you search for a 3 × 3. Even when the 2 × 2 quest fails, the effort has morale value; for next you will search for a 3 × 3 that you *know* is there, which puts an entirely different light on the affair; besides, you've had the fun of drawing pictures.

Probably we should run through a few 3 × m matrixes, just to get used to them. Here is a nice fat one to begin on:

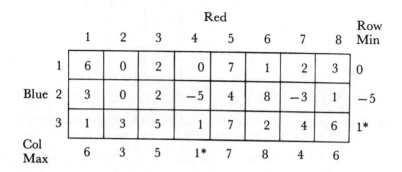

		Red 1	2	3	4	5	6	7	8	Row Min
	1	6	0	2	0	7	1	2	3	0
Blue	2	3	0	2	−5	4	8	−3	1	−5
	3	1	3	5	1	7	2	4	6	1*
Col Max		6	3	5	1*	7	8	4	6	

The usual routine—picking out the row minima and the column maxima, and then the greatest and least of these—has an unusual ending: this monster *does* have a saddle-point! Blue 3 vs. Red 4 are the favored strategies; and the value of the game is 1, to Blue.

You know better than to expect it in the next one:

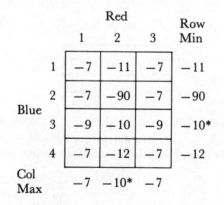

		Red 1	2	3	Row Min
	1	−7	−11	−7	−11
	2	−7	−90	−7	−90
Blue	3	−9	−10	−9	−10*
	4	−7	−12	−7	−12
Col Max		−7	−10*	−7	

But there it is again, which shows you can't trust anyone and you must always look for a saddle-point.

Now consider this game:

	Red 1	2	3	4	5	Row Min
Blue 1	3	5	−2	2	1	−2
Blue 2	3	6	−1	2	4	−1
Blue 3	4	3	6	7	8	3*
Col Max	4*	6	6	7	8	

Here, at last, we find no saddle-point, but a little browsing turns up the fact that Blue 2 is better than Blue 1. This leaves us:

	Red 1	2	3	4	5
Blue 2	3	6	−1	2	4
Blue 3	4	3	6	7	8

Here, Red 3 is better than Red 4 and 5; so we have

	Red 1	2	3
Blue 2	3	6	−1
Blue 3	4	3	6

with which to work. This contains three 2 × 2 games: Red 1, 2; Red 2, 3; and Red 3, 1. The solution to one should solve the original game. It wouldn't cost much to try all of them, but the first one, i.e.,

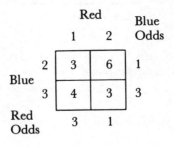

		Red 1	2	Blue Odds
Blue	2	3	6	1
	3	4	3	3
Red Odds		3	1	

happens to serve very well in the original game; so the solution we seek is 0:1:3 for Blue and 3:1:0:0:0 for Red.

Another game:

Red

		1	2	3	4	5
	1	7	1	3	0	2
Blue	2	0	1	6	4	2
	3	1	2	0	5	5

Here, neither saddle-point nor dominance criteria do much for us; so let's hope there is a 2 × 2 solution. The first 2 × 5 subgame

Red

		1	2	3	4	5
	1	7	1	3	0	2
Blue	2	0	1	6	4	2

graphs into

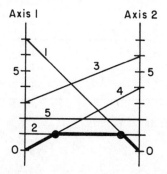

It is clear that this is a somewhat pathological case because of the horizontal line, Red 2. The rule governing graphical solutions tells us to confine our attention to the highest point on the envelope. Here, two intersections, Red 2, 4, and Red 1, 2, vie for this honor. The 2 × 2 subgames, based on Red 2, 4, and Red 1, 2, will have saddle-points, but the 2 × 5 subgame will not; so these saddle-points cannot be solutions to the 2 × 5 subgame. However, the 2 × 2 subgames contain mixed-strategy solutions as well as saddle-points. These are:

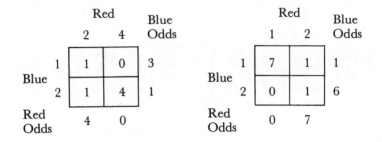

The oddments for Red 1 and 4 in these games turn out to be exactly zero. The interpretation is that Blue should use the 3:1 mix (or the 1:6), while Red plays Red 2; he may also 'play' Red 1 or Red 4, but according to zero oddment! This is of course very confusing. We shall try to return to this point later (much later); just now it isn't worth a digression, since neither of the solutions to this 2 × 5 satisfies the 3 × 5 anyway.

So we try another 2 × 5:

		Red			
	1	2	3	4	5
Blue 2	0	1	6	4	2
Blue 3	1	2	0	5	5

Red 2, 4, and 5 dominate, leaving

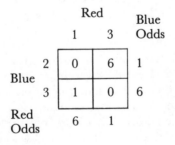

which fails dismally in the original 3 × 5. So we try the remaining 2 × 5:

Red

	1	2	3	4	5
Blue 1	7	1	3	0	2
3	1	2	0	5	5

Red 1 and 5 dominate; so we may graph only Red 2, 3, and 4:

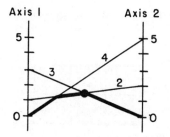

It is clear from this that we want to examine the 2 × 2 based on Red 2, 3:

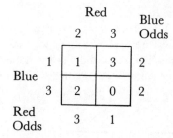

This solves the 2 × 5—as heaven knows it should—but it fails in the original 3 × 5 game.

The enormity of it all may be hard to bear: So far we have learned nothing about this 3 × 5 game, except that it does not have a saddle-point and that it does not have a 2 × 2 solution! The only profit is that we now know it has a 3 × 3 solution. So, back to the mines (i.e., salt mines; the etymology we have in mind goes back to perspiration). Let us try one:

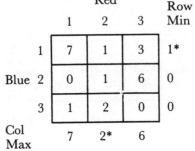

		Red		Row
	1	2	3	Min
1	7	1	3	1*
Blue 2	0	1	6	0
3	1	2	0	0
Col Max	7	2*	6	

No saddle-point, no dominance; so we go after the Red odds, based on

Red

1	2	3
7	0	−3
−1	−1	6

The Red 1 oddment is

0	−3
−1	6

$$= 0 \times 6 - (-3) \times (-1) = -3$$

so 3 is our number. The other Red and Blue oddments are found similarly. For Blue, they are 7:10:32; for Red, 3:39:7—and these also solve the original game. Therefore the work is done (and we have not had to solve the other nine 3 × 3 games which are present in the original matrix):

			Red			Blue
	1	2	3	4	5	Odds
1	7	1	3	0	2	7
Blue 2	0	1	6	4	2	10
3	1	2	0	5	5	32
Red Odds	3	39	7	0	0	

It is customary in scientific writing to state the important points just as often as the unimportant ones: exactly once; and to assume thereafter that this vital information is graven in the minds of all readers. This is one of the wonders of science, for it is economical of space in the journals and it encourages researchers to write up their findings. It also ensures a certain permanence to the work, for it will be years before it is completely understood.

If you are accustomed to that style, then there is an interesting possibility that you are beginning to develop something akin to saddle sores, as a consequence of our ceaseless repetition. You may feel, for instance, that you will curse or cry if you once again encounter an admonition about saddle-points; the sure knowledge that at least one other is being chaffed may help you to endure. As a matter of fact, all (we hope) of this trouble stems from a nightmare on which the writer canters regularly: We picture a poor but honest reader, determined to learn just a little about Game Theory and the solution of small games, who is having a terrible time. He may be sharing a slab of pancake-ice with a polar bear and an empty gun, and may

just be wanting to die intelligently. Or he may be in his study and faced with nothing more lethal than our matrixes, but he is determined to master them. Maybe he wasn't listening the first few times we described a technique or mentioned a concept; or maybe the words used, the times he heard it well, were ambiguous. Whatever his trouble is, the cure is very simple: Just tell it again. We thought you might like to know why you feel that way.

EXAMPLE 19. THE BASS AND THE PROFESSOR

The hero of this story is a bright young centrarchoid (Micropterus), unimaginatively described as 'a perchlike fish much esteemed for food.' The villain is Angler Kleene. These are familiarly known as the Bass and the Professor. Together with certain insects and water, they constitute a natural-habitat group.

The insects—horntails, dragonflies, and bumblebees—are also much esteemed for food. They are not equally common on the surface of the pool; if the Professor adds one (complete with hook) to any species and the Bass feeds on that species, the lethalities are as 2:6:30, which means that there are 5 times as many dragonflies as bumblebees and 3 times as many horntails as dragonflies.* The problem of course is: How should the Bass feed, and the Professor angle? The payoff matrix is this:

		Professor lures with		
		'tails	'flies	'bees
	'tails	−2	0	0
Bass feeds on	'flies	0	−6	0
	'bees	0	0	−30

* Say, $15n$ 'tails, $5n$ 'flies, and n 'bees; and n is large enough so the Professor doesn't upset the frequencies. So the lethalities are like $\frac{1}{15}n : \frac{1}{5}n : \frac{1}{n}$. But n is no friend of the working man; so we multiply through by $30n$, getting 2:6:30. Why do we multiply by $30n$? Well, (a) why not? and (b) to avoid fractions later when he introduces the What's-it, which may be mistaken for any insect, but is twice as likely to arouse suspicion and hence is half as lethal, i.e., 1:3:15.

No saddle-point and no dominance. The shaded matrix for the Bass is

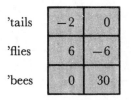

'tails	−2	0
'flies	6	−6
'bees	0	30

so he feeds on horntails according to the oddment

'tails

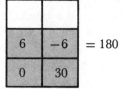

$= 180$

on dragonflies according to

'flies

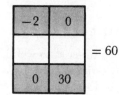

$= 60$

and on bumblebees according to

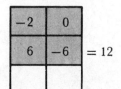

'bees

$= 12$

or, removing the common factors, his grand strategy is 15:5:1. You may verify that the Professor uses the same strategy. The value of the game is negative for the Bass, i.e., -1%.

Now, anglers are great experimentalists and always willing to buy, and possibly use, a new lure, provided it isn't a worm. The Professor purchases a beautiful What's-it, which may be mistaken for any of the three insects; but it is twice as likely to arouse suspicion, which reduces its efficiency. The addition of this changes the game:

Professor

		'tails	'flies	'bees	?
	'tails	-2	0	0	-1
Bass	'flies	0	-6	0	-3
	'bees	0	0	-30	-15

The Bass should now play 3:1:0, which represents a small change in strategy—the bees are now too dangerous to eat. The Professor should play 7:2:0:1, and should never use the bee lure. His new lure, in fact, plays only 10 per cent of the time. The value of the game is now $-3\%_0$, which is a very slight improvement, for the Professor, over the $-3\%_1$ value of the original game. This confirms a well-known empirical finding—that fishing is expensive.

EXAMPLE 20. THE BEDSIDE MANNER

The patient is suffering from a well-known disease; so well known, in fact, that five variants, identified with five strains of bacteria, have been observed. Dr. Wendel is quite unable to pin it down from a bedside diagnosis.

He again feels for the pulse. The patient's confidence goes up one notch, for this recheck is evidence of a careful, conscientious practitioner. The subterfuge would fail and confidence would fall, noisily, if he knew the doctor's reason for the recheck:

Dr. Wendel was manufacturing an excuse to look at his second hand; he needed a random number, of course. Which shows that the bases of confidence are very tricky indeed, for the doctor was being super-conscientious and fully deserved the patient's confidence; he just didn't have time to teach him both Medicine and Game Theory.

The doctor has three medicines: The first has a fifty-fifty chance of overcoming four of the bacteria strains; the second is lethal to one of these four; the third is a fifty-fifty bet against another of these, and it is sure-fire against the fifth strain.

The matrix is:

	Strain 1	2	3	4	5
Medicine 1	½	½	½	½	0
Medicine 2	1	0	0	0	0
Medicine 3	0	½	0	0	1

From Nature's point of view, Strains 1 and 2 dominate Strains 3 and 4, and the latter two are equivalent against this battery of medicines; so the game comes down to this:

	Strain 4	5
Medicine 1	½	0
Medicine 2	0	0
Medicine 3	0	1

In this subgame, it doesn't require genius on the part of the doctor to discern that Medicine 2 is dominated by the others; so we have:

	Strain 4	5	Doctor Odds
Medicine 1	½	0	1
Medicine 3	0	1	½
Nature Odds	1	½	

The odds are 1:½ or 2:1, in favor of Medicine 1; so the doctor's mixed strategy is 2:0:1. The value of the game is ⅓, and Nature cannot maneuver his success average below that value.

It may be worth making a short digression at this point to show how the situation changes if the doctor knows something about Nature's strategy. Since She is not really malevolent, She will not go out of Her way to punish him for using his knowledge.

Suppose that experience indicates that Nature mixes these strains according to the odds 1:3:3:2:5. The doctor can then estimate the average return from each of his strategies. For example, if he consistently uses Blue 1, his average winnings will be (from the payoffs in the original matrix)

$$\frac{1 \times \frac{1}{2} + 3 \times \frac{1}{2} + 3 \times \frac{1}{2} + 2 \times \frac{1}{2} + 5 \times 0}{1 + 3 + 3 + 2 + 5} = \frac{\frac{9}{2}}{14} = \frac{9}{28}$$

His average when using Blue 2 will be

$$\frac{1 \times 1 + 3 \times 0 + 3 \times 0 + 2 \times 0 + 5 \times 0}{14} = \frac{1}{14} \left(= \frac{2}{28} \right)$$

For Blue 3 it turns out to be $\frac{13}{28}$. So the game now looks like this:

Nature

Doctor 1		$\frac{9}{28}$
2		$\frac{2}{28}$
3		$\frac{13}{28}$

and it is clear that he should use Blue 3. The value of this game is $\frac{13}{28}$, which is greater than the value ($\frac{1}{3}$) of the original game. Thus the additional information that we are now assuming he possesses turns out to be worth something to him. That is generally true, especially in games against Nature: The more you know about Her, the better you can play against Her.

The discussion can profitably be carried one step further: If Nature is really using the grand strategy 1:3:3:2:5, as we now assume, and the doctor in his ignorance uses the grand strategy 2:0:1, which is prescribed by Game Theory, he will in fact win more than ⅓, which is the value of the original game. For Nature is not playing her best strategy. We have seen that Her use of 1:3:3:2:5 reduces the game to

Nature

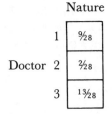

Doctor 1	$\frac{9}{28}$
2	$\frac{2}{28}$
3	$\frac{13}{28}$

If he uses the 2:0:1-mix—which was suggested by his original analysis —in this game, his average gain will be

$$\frac{2 \times \frac{9}{28} + 0 \times \frac{2}{28} + 1 \times \frac{13}{28}}{2 + 0 + 1} = \frac{31}{84}$$

We may summarize as follows: If the game is as the doctor sees it, and Nature does Her worst, he will win $\frac{1}{3} = 0.333$. If Nature is playing 1:3:3:2:5, he will win $\frac{31}{84} = 0.369$. If he knows Her strategy, he can change his from 2:0:1 to 0:0:1 and win $\frac{13}{28} = 0.464$; so this extra knowledge is worth about 26 per cent to him.

There must be many real situations in which partial information and Game Theory technique will enable one to give a good account of himself, even though he may fall short of the value of the real game.

EXAMPLE 21. THE CHESSERS

Two three-man teams forgather to play chess; two games. The event is known as Barankin-Bodenhorn-Brown vs. Bellman-Brown-Brown. We regret that there are so many Browns, but that's life; however, identifications are actually unique, as will be clear to all good puzzle solvers.

Brown can beat Barankin, Barankin can beat Bellman, and Bellman can beat Bodenhorn; otherwise the players are equal. Brown becomes ill at the last moment. This information is summarized, for convenience:

	Ba	Bo	Br
Be	−1	1	0
Br	1	0	0

Each team must select two players and the order in which they will play. To compensate somewhat for the loss through illness, it is agreed that team captain Brown may play both games for his reduced team, if he desires.

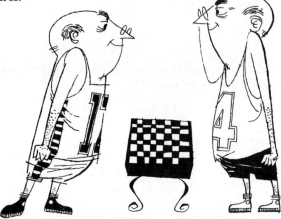

The possibilities are these (where BaBo indicates that Barankin plays in the first game, Bodenhorn in the second, etc.):

	(1) BaBo	(2) BaBr	(3) BoBa	(4) BoBr	(5) BrBa	(6) BrBo
BeBr	−1	−1	2	1	1	0
BrBe	2	1	−1	0	−1	1
BrBr	1	1	1	0	1	0

There are no saddle-points, but (1) dominates (2), and (3) dominates (5); so the game reduces to:

(2)	(4)	(5)	(6)
−1	1	1	0
1	0	−1	1
1	0	1	0

Just a little exploration will lead you to the solutions: the reduced team should mix its strategies according to the odds 1:1:1. There are two basic solutions which the other team may use: 0:1:0:2:0:0 or 0:0:0:0:1:2.

SUMMARY OF 3 × m METHODS

The methods for solving $3 \times m$ games differ only a little from those used on 3×3's. You begin, as usual, by searching for a saddle-point and, if that quest fails, continue by trying to eliminate some rows and columns by dominance arguments. If this does not reduce the game to something you can handle—i.e., to a 2×2, a $2 \times m$, or a 3×3—then you may proceed further by either of two routes:

1. Choose some 3×3 subgame, solve it, and test this solution in the original $3 \times m$ game. Continue this process until you succeed.
2. Break the game up into three $2 \times m$ subgames—based on Blue 1, 2, Blue 1, 3, and Blue 2, 3, respectively—and, by the graphical method, search for a 2×2 subgame which satisfies the $2 \times m$ and (hopefully) the original $3 \times m$ game. If this fails, there is no 2×2 solution, and you must follow route 1.

The following Exercises are the best ones yet; it is the first set on which your suffering will probably equal ours.

EXERCISES 4

Determine the oddments and values of the following games:

1.

Red

		1	2	3	4
	1	1	7	3	4
Blue	2	5	6	4	5
	3	7	2	0	3

2.

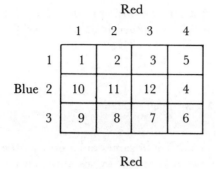

Red

		1	2	3	4
	1	1	2	3	5
Blue	2	10	11	12	4
	3	9	8	7	6

3.

Red

		1	2	3	4
	1	1	6	2	5
Blue	2	5	1	6	2
	3	2	5	1	6

4.

Red

		1	2	3	4
	1	6	0	1	2
Blue	2	0	3	1	0
	3	2	0	3	1

5.

Red

		1	2	3	4
	1	1	0	1	0
Blue	2	0	1	0	1
	3	1	0	0	1

6.

Red

		1	2	3	4
	1	0	1	2	3
Blue	2	3	2	1	0
	3	0	2	1	3

7.

Red

		1	2	3	4
	1	8	0	6	7
Blue	2	3	3	4	8
	3	3	6	3	1

8.

Red

		1	2	3	4
	1	3	4	0	2
Blue	2	6	1	3	0
	3	2	5	3	6

9.

Red

		1	2	3	4	5
	1	1	3	5	3	1
Blue	2	4	2	0	0	2
	3	1	2	1	4	4

10.

Red

		1	2	3	4	5	6
	1	4	3	3	2	2	6
Blue	2	6	0	4	2	6	2
	3	0	7	3	6	2	2

11.

Red

		1	2	3
	1	8	2	−4
	2	4	1	−2
Blue	3	0	0	0
	4	−4	−1	2
	5	−8	−2	4

12.

Red

		1	2	3
	1	0	−17	−34
	2	−2	−15	−35
	3	−20	−22	−24
Blue	4	−3	−15	−35
	5	−40	−27	−14
	6	−5	−21	−30

13.

Red

		1	2	3
	1	−1	0	1
	2	1	0	0
	3	1	−1	0
Blue	4	5	−4	−2
	5	0	0	1
	6	0	1	0
	7	25	−13	−12

14.

Red

		1	2	3
	1	3	5	1
Blue	2	−6	2	0
	3	5	4	3
	4	−7	3	6

15.

Red

		1	2	3
	1	3	1	4
Blue	2	1	5	9
	3	2	6	5
	4	3	5	8

Four-strategy Games and Larger Ones

You will find that solving 4 × 4 games, and larger ones, is not a beautiful experience, unless your affinity for arithmetic is clinically interesting. However, the game situations compensate for this, to some extent, by being more interesting; and one's feeling of well-being is quite marked *after* solving a large game.

SOLUTION VIA REVELATION

The best way to get the solution to a large game is through Revelation.* If you can guess the solution from the structure of the problem, or obtain a promising suggestion from any source, test it: simply calculate the average payoff when Blue, say, uses his presumed-good strategy against each (in turn) of Red's pure strategies. If you have actually guessed the solution, then two things should be so: First, these payoffs should all be equal, except, possibly, for larger payoffs that Blue may earn against pure strategies that are not used in Red's best mixture. Second, this average payoff should appear again when Red plays his appropriate mixed strategy against each Blue pure strategy; but the payoff may be smaller (and hence better for Red) against a Blue strategy which is not involved in Blue's good mixture. An example will show how much easier it is to say this than to do it.

Consider this 5 × 5 game:

			Red		
	1	2	3	4	5
1	0	1	0	2	0
2	3	0	0	1	2
Blue 3	0	0	2	1	2
4	1	3	0	0	1
5	1	2	3	1	0

If you are the seventh son of a seventh son, have a crystal ball of un-

*The reader who intends to solve large games should turn to Chapter 6 about now to learn a powerful method.

usual clarity, and are fortunate, it may cross your mind that Red should mix his five strategies according to the odds 4:35:6:57:40. (This is a very complicated set of numbers, quite beyond the modest powers we impute to you above; but the example emphasizes that you can—and should—test any solution, or presumed solution, and satisfy yourself that it is, or is not, valid.)

The pertinence of these numbers, if any, may be settled by assuming, tentatively, that they are true, and calculating the value of the game. For convenience, write:

4	35	6	57	40
0	1	0	2	0
3	0	0	1	2
0	0	2	1	2
1	3	0	0	1
1	2	3	1	0

Now, taking one row at a time, multiply the oddments by the corresponding payoffs, add all of these together, and divide by the sum of the oddments. Thus, against the first Blue strategy (first row), the average payoff is

$$\frac{4 \times 0 + 35 \times 1 + 6 \times 0 + 57 \times 2 + 40 \times 0}{4 + 35 + 6 + 57 + 40} = \frac{149}{142}$$

Against the second Blue strategy, it is (leaving out the things multiplied by zero)

$$\frac{4 \times 3 + 57 \times 1 + 40 \times 2}{142} = \frac{149}{142}$$

against Blue 3,

$$\frac{6 \times 2 + 57 \times 1 + 40 \times 2}{142} = \frac{149}{142}$$

against Blue 4,

$$\frac{4 \times 1 + 35 \times 3 + 40 \times 1}{142} = \frac{149}{142}$$

and against Blue 5,

$$\frac{4 \times 1 + 35 \times 2 + 6 \times 3 + 57 \times 1}{142} = \frac{149}{142}$$

It is now clear that the source of your information regarding Red's mixture is one you should cherish, for the steady-state result of its use by Red—a loss of $\frac{149}{142}$ payoff units—appears pregnant with meaning. However, we have not proved that these are the best odds; it may be that Blue's best mix will ensure that he wins at least $\frac{148}{142}$, say, rather than $\frac{149}{142}$; so we must also have the appropriate odds for Blue. If you can get the above set of oddments for Red, it imposes little additional strain to suppose you also have access to these for Blue: 28:33:31:21:29. You may then verify that the payoff to Blue is $\frac{149}{142}$.

Every game solution should be verified in this way. It is your sole insurance against frantic arithmetic, and other inadequacies of man or method.

SADDLE-POINTS

If a large game has one or more saddle-points, the solution is easy to find. For example, consider:

		Red			Row Min
	1	2	3	4	
Blue 1	6	5	6	5	5*
2	1	4	2	−1	−1
3	8	5	7	5	5*
4	0	2	6	2	0
Col Max	8	5*	7	5*	

The largest row minimum is 5 and the smallest column maximum is 5; so there is a saddle-point. There are four of them, in fact. (Note that *when there are several saddle-points, all have the same value.*) So Blue may play the pure strategy Blue 1, or he may play Blue 3. Red may play either Red 2 or Red 4; he may mix these two, if he wishes, in any way. The value of the game is 5.

DOMINANCE

Just as in the smaller games, it pays to look for dominating columns and dominated rows. Once found, they should be stricken out of the game matrix or assigned zero oddments. For example, the following game is only a 3 × 3, in effect. Each unmarked row and column is less desirable than some marked (†) one, from the respective viewpoints of the two players.   Red

	1	2	3	4	5	6	
1	3	5	7	2	3	6	
2	7	5	5	4	5	5	†
3	4	6	8	3	4	7	†
4	5	0	3	4	4	2	
5	7	2	2	3	5	3	
6	6	3	4	5	6	5	†
		†		†	†		

(Blue, on the left margin)

It is probably clear from this example that searching for dominance in a large game matrix, while simple in principle, is likely to be an eyesoring operation.

ALL-STRATEGIES-ACTIVE

As in the case of 3 × 3 games, after testing for saddle-points and dominance, we make the tentative assumption that all strategies are active in the best mixed strategy, and try to compute that mix. To do so, we must adjoin to the 3 × 3 technique one further technical trick. We shall describe the process through a particular 4 × 4 game, namely,   Red

	1	2	3	4
1	1	7	0	3
2	0	0	3	5
3	1	2	4	1
4	6	0	2	0

(Blue, on the left margin)

but its generalization to games of any size will be self-evident.

As usual, we begin by considering the problem for Red. First, subtract from each payoff the payoff directly below it—just as we did in the 2 × 2 and 3 × 3 games. This gives us

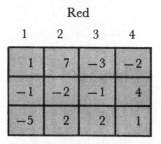

Red

1	2	3	4
1	7	−3	−2
−1	−2	−1	4
−5	2	2	1

The oddment for any particular pure strategy is based on what is left after the column identified with that strategy is stricken from this shaded matrix. Thus the oddment for Red 1, for example, is determined from these numbers:

1

	7	−3	−2
	−2	−1	4
	2	2	1

The process of coaxing the oddment out of these shaded arrays is easy, in principle, even for large arrays, just as it is easy to count to a million by ones; at each step you will know exactly what to do next. Also, it is easier to do than to explain; unfortunately, we must jointly undertake an explanation.

In order to determine the oddment implicit in any shaded array, we must get some zeros mixed in among the other numbers. In particular, we must *convert all but one of the numbers in some row (or in some column) to zeros*, using some legitimate method, of course. Thus, we aspire to convert the 3 × 3 array

7	−3	−2
−2	−1	4
2	2	1

into, say,

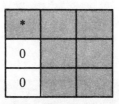

or perhaps into

where the shaded boxes contain suitable numbers, probably different from those in the original. The asterisked box is so marked just to identify the nonzero element in the column or row. These remarks apply to larger matrixes too; thus if we had one that looked like this:

we should be pleased to get it changed to this:

Why all this interest and pleasure in zeros? Strictly utilitarian of course; for, *once this form is reached, we may disregard all boxes that lie due east, due west, due north, and due south of the asterisk,* and the asterisked box itself may be lifted out of the matrix as a factor. Thus

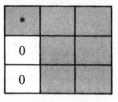

is equivalent to

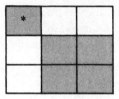

which is equivalent to

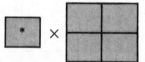

Similarly,

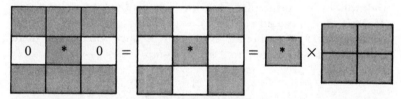

(In the final step here, the outlying boxes just close up their ranks.)

This process immediately reduces a 3 × 3 array to a 2 × 2 array, multiplied by the number in the asterisked box. We know how to find the value of 2 × 2 arrays—from the difference of the products of the diagonal boxes. In the case of a 4 × 4 array, the process must be applied twice in order to reduce it to a 2 × 2 array (and two multipliers).*

---

* This process is neither as silly nor as fortuitous as it may seem to you on first reading. Actually there is a straightforward mathematical theory which leads to and justifies it. If you ever learned, in a course in algebra, how to evaluate a determinant, then you will have recognized this process long before reading this footnote.

Having first convinced you, we trust, of the desirability of having strategically placed zeros in the shaded array, we now turn to means of producing them. We are guided in this by the following house rules:

1. You may add the numbers in any row to the numbers in any other row—on a box-by-box basis.
2. Subtraction has the same social standing as addition, so you can subtract in the same box-by-box way.
3. You may multiply the numbers of a row by anything you want before adding them to another row.
4. Columns are as good as rows, so the foregoing statements about things you can do to rows apply equally well to columns.
5. Be artful about it.

The force of Rule 5 is that it is easier to make zeros in some places than in others; so look around for a likely point to attack.

Let us return now to

7	−3	−2
−2	−1	4
2	2	1

and apply these principles. The last row contains two 2's and a 1 (Rule 5). By subtraction (Rule 2), with small quantities of multiplication (Rule 3), we should be able to produce some zeros there. Begin by subtracting column 2 from column 1:

10	−3	−2
−1	−1	4
0	2	1

This produces one zero. We can make another alongside it by multiplying the last column by 2 before subtracting it from the second column; this yields

10	1	−2
−1	−9	4
0	0	1*

which is just the sort of form we hoped to get, for a row now has zeros in every box but one. But

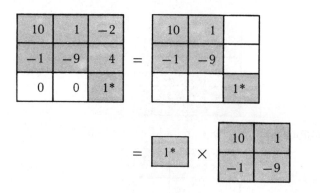

$$= 1 \times [10 \times (-9) - 1 \times (-1)] = -89$$

so 89 is the oddment for Red 1 in the game on page 136. If you are particularly proud of your insight in these matters, you are in a dandy position to try it out now for oddment Red 2, before making the calculation.

Striking out the second column in the original shaded matrix, i.e., in

Red

	1	2	3	4
1	7	−3	−2
−1	−2	−1	4
−5	2	2	1

we have

1	−3	−2
−1	−1	4
−5	2	1

which governs the oddment Red 2. The second row looks promising for zeros; so we begin by subtracting column 2 from column 1, which yields

4	−3	−2
0	−1	4
−7	2	1

We then multiply column 2 by 4 and add it to column 3:

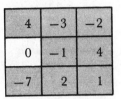

4	−3	−14
0	−1*	0
−7	2	9

This is equivalent to

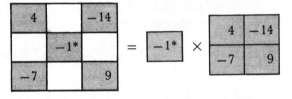

$$= -1 \times [4 \times 9 - (-7) \times (-14)] = 62$$

the oddment Red 2. Similarly, the oddments Red 3 and 4 are found to be 119 and 83.

To determine Blue's grand strategy, we must return to the original game matrix, i.e., to

		1	7	0	3
	1	1	7	0	3
	2	0	0	3	5
Blue	3	1	2	4	1
	4	6	0	2	0

and form a new shaded matrix by subtracting each payoff from its left-hand neighbor:

<table>
<tr><td>1</td><td>−6</td><td>7</td><td>−3</td></tr>
<tr><td>2</td><td>0</td><td>−3</td><td>−2</td></tr>
<tr><td>Blue   3</td><td>−1</td><td>−2</td><td>3</td></tr>
<tr><td>4</td><td>6</td><td>−2</td><td>2</td></tr>
</table>

Striking out the first row, we have

<table>
<tr><td>0</td><td>−3</td><td>−2</td></tr>
<tr><td>−1</td><td>−2</td><td>3</td></tr>
<tr><td>6</td><td>−2</td><td>2</td></tr>
</table>

which governs the oddment Blue 1. The zeros necessary for its evaluation may be obtained by adding 6 times row 2 to row 3, which results in

<table>
<tr><td>0</td><td>−3</td><td>−2</td><td rowspan="3">=</td><td></td><td>−3</td><td>−2</td></tr>
<tr><td>−1*</td><td>−2</td><td>3</td><td>−1*</td><td></td><td></td></tr>
<tr><td>0</td><td>−14</td><td>20</td><td></td><td>−14</td><td>20</td></tr>
</table>

$$= \boxed{-1^*} \times \begin{array}{|c|c|} \hline -3 & -2 \\ \hline -14 & 20 \\ \hline \end{array}$$

$$= -1 \times [(-3) \times 20 - (-14) \times (-2)] = 88$$

The oddments Blue 2, 3, and 4 are found similarly. The ultimate results are summarized here:

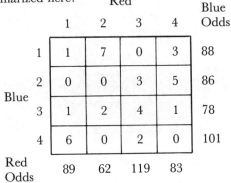

		Red			Blue
	1	2	3	4	Odds
1	1	7	0	3	88
2	0	0	3	5	86
Blue 3	1	2	4	1	78
4	6	0	2	0	101
Red Odds	89	62	119	83	

It is of course necessary to verify that these odds do constitute a solution; we have done so, and you may wish to do so. The value of the game may be found by using Red's best mix against Blue 2, say (of course the best mix of either player may be used against any pure strategy in the other's best mix), i.e.,

$$\frac{119 \times 3 + 83 \times 5}{89 + 62 + 119 + 83} = \frac{772}{353}$$

so Blue's winnings, with good play, will average a little more than 2 units per play.

Before leaving you standing, paralyzed, before the next set of exercises, we doubtless should conduct you through one example which requires a double application of the method of reducing matrixes. In the game

			Red		
	1	2	3	4	5
1	0	1	0	2	0
2	3	0	0	1	2
Blue 3	0	0	2	1	2
4	1	3	0	0	1
5	1	2	3	1	0

we leave it to you to verify that there are neither saddle-points nor dominant strategies (or are there?). Then, attending to Red's mix, we subtract from each payoff the payoff directly below it. This gives us:

Red

1	2	3	4	5
−3	1	0	1	−2
3	0	−2	0	0
−1	−3	2	1	1
0	1	−3	−1	1

Suppose now, for the sake of an illustration, that we wish to find the oddment Red 3. Striking out the Red 3 column, the pertinent remaining matrix turns out to be

−3	1	1	−2
3*	0	0	0
−1	−3	1	1
0	1	−1	1

Row 2 happens to be in the form we require—all entries are zero, except for one—so we may reduce the matrix immediately, beginning by casting out all entries which are due north, south, or east of 3 * :

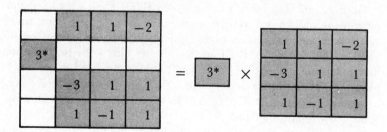

To reduce this 3×3, we are attracted by the numbers in column 2; it is evident that we can get the desired zeros by adding row 3 to row 1, then to row 2. Doing so, we get

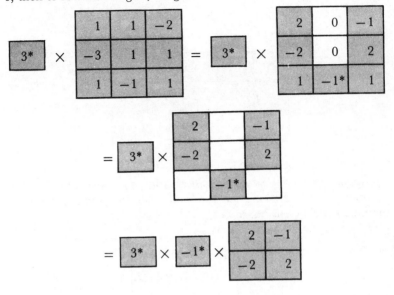

$$= 3 \times (-1) \times [2 \times 2 - (-2) \times (-1)] = -6$$

You may perhaps recognize this 6 as oddment Red 3, for this example is the one you solved by Revelation at the beginning of the chapter.

EXAMPLE 22. THE SECONDHAND CAR

Gladyn and Don inherit a car worth $800. The evils of communism being well known to them, they agree to settle the ownership by means of sealed bids. The high bidder gets the car by paying his brother the amount of the high bid. If the bids are equal—which they may well be, because they agree to bid in hundred-dollar quantities—the ownership is determined by the toss of a coin, there being no exchange of funds. Gladyn has $500 on hand, whereas Don has $800. How should they bid?

In the event that ownership is decided by the toss of a coin, their

expectations are equal. Since the car is worth 800, this expectation is therefore 400. With this information at hand, we may now fill in the payoff matrix showing Gladyn's expected gains (in hundreds of dollars).

		Don's bid									Row Min
		0	1	2	3	4	5	6	7	8	
	0	4	1	2	3	4	5	6	7	8	1
	1	7	4	2	3	4	5	6	7	8	2
Gladyn's bid	2	6	6	4	3	4	5	6	7	8	3
	3	5	5	5	4	4	5	6	7	8	4*
	4	4	4	4	4	4	5	6	7	8	4*
	5	3	3	3	3	3	4	6	7	8	3
Col Max		7	6	5	4*	4*	5	6		8	

We test this for a saddle-point and find immediately that it has one; because the minmax and the maxmin are both equal to 400. In fact the game has four saddle-points: each player should bid either 300 or 400. Actually, the 300 bid is somewhat more attractive, because it will inflict greater penalties if the other brother does not bid well.

EXAMPLE 23. THE SILVICULTURISTS

The tree-spraying firm of Harris and Danskin was dissolving. The partners wished to buy from the company its two capital items: an air compressor worth $400 and a quarter-ton pickup worth $600.

They decided to submit sealed bids, in multiples of $200, for each item. In the event of a tie, the buyer would be decided by a hand of showdown Poker. The proceeds of the sales would be assets of the company and therefore divided between the partners.

We shall represent a bidding strategy by a two-digit index, such as 42, which means a bid of $400 for the compressor and $200 for the truck. The payoff matrix (to Harris), in hundreds of dollars, is then easily calculated. For example, if Harris bids 22 and Danskin bids 06,

Harris will receive the compressor, worth 4, and one-half of the cash assets, 2 + 6, worth 4; so, having paid in 2, he nets 6. Ties are figured at one-half the nominal values, to reflect the vagaries of Poker. The matrix is:

		00	02	04	06	20	Danskin 22	24	26	40	42	44	46	Row Min
	00	5	3	4	5	4	2	3	4	5	3	4	5	2
	02	7	5	4	5	6	4	3	4	7	5	4	5	3
	04	6	6	5	5	5	5	4	4	6	6	5	5	4
	06	5	5	5	5	4	4	4	4	5	5	5	5	4
	20	6	4	5	6	5	3	4	5	5	3	4	5	3
	22	8	6	5	6	7	5	4	5	7	5	4	5	4
Harris	24	7	7	6	6	6	6	5	5	6	6	5	5	5*
	26	6	6	6	6	5	5	5	5	5	. 5	5	5	5*
	40	5	3	4	5	5	3	4	5	5	3	4	5	3
	42	7	5	4	5	7	5	4	5	7	5	4	5	4
	44	6	6	5	5	6	6	5	5	6	6	5	5	5*
	46	5	5	5	5	5	5	5	5	5	5	5	5	5*
Col Max		8	7	6	6	7	6	5*	5*	7	6	5*	5*	

It is quickly established that this 12 × 12 game is rife with saddle-points—16 of them, corresponding to the strategies 24, 26, 44, and 46. Any of these will ensure a fair chance to each partner, but some are more punishing than others, in the presence of poor play. Bids of 200 and 400, for the compressor and truck, respectively, may net Harris 700 if Danskin is careless. If Danskin makes these bids, Harris may net as little as 300. If they don't play the saddle-points, the payoff can range from 200 to 800.

Since the value of the game is $500 to Harris, it may appear to be an unfair game to Danskin. Actually, it is symmetric: the firm is worth $1000 and Danskin will get half of it.

The remarks in this example and in the last one, to the effect that one good strategy may at times be better than another good one, are sufficient to kindle an argument in any representative group of Game Theorists. The strict priests of the cult will protest that the minimax principle—the equality of the maxmin and minmax, in the case of a saddle-point—is the sole guide and criterion in the body of doctrine known as Game Theory. And further, that so long as that criterion is satisfied, nothing more need be said. The application of this viewpoint to the present situation is this: We arrive at a saddle-point by assuming that the enemy is smart enough to appreciate its merits. Having accepted this viewpoint, let's not turn around (they say) and argue that one saddle-point is better than another because the enemy may not play a saddle-point strategy.

The writer has some sympathy for this attitude, but not much.

EXAMPLE 24. COLOR POKER

Mel and Ray cannot count reliably, which somewhat handicaps them in conventional Poker. They are good at colors, however, so they devise this variant: Each antes $1, after which one card is dealt to each from a shuffled deck; that is all they get. A black card is better than a red card; otherwise they all look alike to these peasants.

Mel, the first player, may *pass,* or *bet* $1. If he passes, a showdown occurs, and the pot goes to the high-card holder (or is split if the cards are equal). If he bets, then Ray may *fold* (thereby losing), *see* by betting $1 (thereby forcing a showdown), or *raise* by betting $2.

If Ray raises, Mel may *fold* (thereby losing), or he may *see* by betting $1 (followed by showdown).

Thus a player may risk up to a total of $1, $2, or $3 in each pot, and his strategies are defined by the various totals he is willing to risk on a red card and on a black card. Thus (1, 3) will indicate a strategy in which a player is willing to risk up to $1 on a red card and up to $3 on a black one. There are nine strategies of this sort open to the players.

Setting up the payoff matrix is easy, but tedious. For each pair of strategies we make a calculation of this type [we use Mel (2, 2) vs. Ray (3, 2) for illustration]:

Ray holds

	Red and bets 3	Black and bets 2
Mel holds Red and bets 2	−2	−2
Black and bets 2	−2	0

The two left boxes correspond to cases where Mel loses $2; he is bluffed out of the game by Ray's high bids. The upper right box represents a loss of $2 in a showdown against a better hand; the lower right, equal hands and a draw. The deck contains equal numbers of red and black cards, so the odds favoring the four events, to which the boxes correspond, are equal. Therefore the average value of Mel (2, 2) vs. Ray (3, 2) is

$$\frac{1 \times (-2) + 1 \times (-2) + 1 \times (-2) + 1 \times 0}{1 + 1 + 1 + 1} = -\frac{6}{4}$$

a loss for Mel of 6 quarters. Since the betting is in dollars, these quarters are in fact 25-cent pieces. It is convenient to use them in the pay-off matrix, so in the matrix just below an entry indicates the numbers of 25-cent pieces that change hands (on the average).

	Ray 1,1	1,2	1,3	2,1	2,2	2,3	3,1	3,2	3,3	Row Min
1,1	0	0	0	0	0	0	0	0	0	0*
1,2	1	0	−2	2	1	−1	−2	−3	−5	−5
1,3	1	0	0	2	1	1	3	2	2	0*
2,1	3	0	0	2	−1	−1	0	−3	−3	−3
2,2	4	0	−2	4	0	−2	−2	−6	−8	−8
2,3	4	0	0	4	0	0	3	−1	−1	−1
3,1	3	0	−1	2	−1	−2	2	−1	−2	−2
3,2	4	0	−3	4	0	−3	0	−4	−7	−7
3,3	4	0	−1	4	0	−1	5	1	0	−1
Col Max	4	0*	0*	4	1	1	5	2	2	

(Mel labels the rows.)

The analysis, happily, is not tedious: At the outset we find saddle-points—four of them—corresponding to Mel (1, 1) or (1, 3) against Ray (1, 2) or (1, 3). Thus neither player should bluff, in the sense of bidding high on a poor hand, but Mel may make disarming low bids (and Ray medium ones) on good hands.

The value at the saddle is zero, so the game is fair. This is somewhat unexpected, in view of the lack of symmetry in the play and in
the payoff matrix. This example gives you a glimpse of the complica-
tions which would arise if we were to analyze conventional Poker.

EXAMPLE 25. FOR OLDER CHILDREN

In Chapter 3 we discussed a children's game called Scissors-Paper-
Stone. This may be generalized, in two fundamentally different ways,
to a five-element game for older children. One of these has the in-
tuitive solution, and the other does not.

Let's call this game Stone-Water-Scissors-Glass-Paper. It is obvious
to any child that Water wets Stone and Paper, Scissors cost more than
Water and Stone, Glass is more brittle than Water and Scissors,
Paper is more flexible than Scissors and Glass, and Stone is thicker
than Glass and Paper. These relations may be diagrammed as follows,
where the arrows indicate the direction of flow of assets:

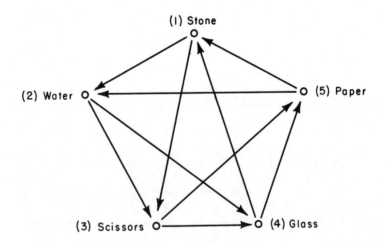

If we count win, lose, and draw for Blue as 1, −1, and 0, respectively, and if we identify the strategies by the numbers in the diagram, the payoff matrix becomes:

Red

		1	2	3	4	5
	1	0	−1	−1	1	1
	2	1	0	−1	−1	1
Blue	3	1	1	0	−1	−1
	4	−1	1	1	0	−1
	5	−1	−1	1	1	0

After testing, fruitlessly, for saddle-points and dominance, the all-strategies-active method finally yields the solution, namely, 1:1:1:1:1 for each player; i.e., they should mix their strategies equally.

If the directions of three arrows are reversed—say by deciding that after all Paper burns better than Water, that Scissors cut Paper, and that Stone breaks Scissors—i.e.,

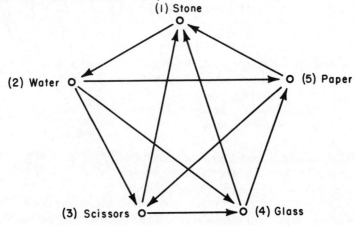

(1) Stone

(2) Water   (5) Paper

(3) Scissors   (4) Glass

the payoff matrix becomes   Red

		1	2	3	4	5
	1	0	−1	1	1	1
	2	1	0	−1	−1	−1
Blue	3	−1	1	0	−1	1
	4	−1	1	1	0	−1
	5	−1	1	−1	1	0

After a brief return to the mines, you will find that the solution for Stone-Water-Scissors-Glass-Paper is now 3:3:1:1:1.

EXAMPLE 26. THE PROCESS SERVER

Gary, a process server, watches first one door and then the other, hoping to catch Steve when he comes out. Steve's friend, Scott, is in the building, too, and is willing to assist Steve. Never having had this trouble before, they cannot identify Gary in the crowd.

They may leave by either door and either may go first. Gary may tag either the first or second man who appears at the door he is watching, but if he tries to serve Scott while Steve is watching from a window, Steve will escape through the other exit. A hit on Steve counts as 1 for the law and a miss as 0.

Gary has four pure strategies: he may watch either door and tag or pass up the first man he sees.

Steve has a richer—and more complicated—collection: Two of his strategies are to go first, through either door; no further decisions are required if he uses one of these. He has four more strategies, one of which is to send Scott out one door and watch for developments; if Scott is tagged, to escape through the other door, and if Scott is not tagged, to *follow* him—and hope for the best. There are four strategies of this type, obtained by sending Scott out either the front or rear door and by following or not following him through the same door.

We have a 4 × 6 game. The pure strategies having been unraveled, it is very easy to write the payoff matrix. The code is: *F* or *R* denotes Steve's use of front or rear door; *f* or *r* denotes the same for Scott.

Steve and Scott

Door Watched	Man Tagged		*F* (1)	*R* (2)	*f,F* (3)	*f,R* (4)	*r,R* (5)	*r,F* (6)
Front	1st	(1)	1	0	0	0	0	1
	2nd	(2)	0	0	1	0	0	0
Rear	1st	(3)	0	1	0	1	0	0
	2nd	(4)	0	0	0	0	1	0

Our usual methods of solution work very well, as there are 4 active strategies in the good mixed strategies. There are several basic mixed strategies in this game: Steve may play the odds 1:1:1:1 with any 4 pure strategies which form a 4 × 4 matrix having a 1 in each row and column; for example, strategies 3, 4, 5, 6.

Gary's best mix is 1:1:1:1, and the value of the game is ¼ (to Gary). If Steve were alone in the building, the value would be ½; so Scott's connivance is very worth while.

EXAMPLE 27. THE PALM GAME

This game involves two players, each of whom palms a coin, or doesn't, after which each guesses the total number palmed. The strategies available to a player are:

(*a*) Palm no coin, guess total is 0.
(*b*) Palm no coin, guess total is 1.
(*c*) Palm 1 coin, guess total is 1.
(*d*) Palm 1 coin, guess total is 2.

Counting ties as 0, wins and losses for Blue as $+1$ and -1, we get the following game:

Red

		1	2	3	4
	1	0	1	-1	0
Blue	2	-1	0	0	1
	3	1	0	0	-1
	4	0	-1	1	0

This game has no saddle-point and no dominant strategies, and our all-strategies-active method results in zero odds for all concerned; so we have troubles: there must be a smaller game, embedded in the 4×4, whose solution satisfies the larger game. Among the 2×4 subgames is

Red

		1	2	3	4
Blue	2	-1	0	0	1
	3	1	0	0	-1

It is easy to establish that Blue 2 and Blue 3 should be played according to the odds 1:1 and that Red should play1:0:0:1; and this solution holds up in the original game.

However, observe that this grand strategy is not the sole one available; for this 2×4,

Red

		1	2	3	4
Blue	1	0	1	-1	0
	4	0	-1	1	0

also has a solution which satisfies the original game, namely, 1:0:0:1 and 0:1:1:0 for Blue and Red, respectively.

So now we know two good ways to play the game: Use strategies 1 and 4, or use strategies 2 and 3; and whichever pair you use, mix them by an equal-odds device. It may also be clear that we can go a step further and get a more general solution based on both of these grand strategies: namely, use all four strategies mixed any way you like, provided the oddment for Blue 1 is the same as for Blue 4 and the oddment for Blue 2 is the same as for Blue 3. (The same for Red.) Thus the odds 7:3:3:7 constitute a good mixed strategy, for either player, as do the odds 3:7:7:3, or any other similar set of numbers.

This is the first time that we have emphasized the phenomenon that a game may have more than one good grand strategy; but we have seen it before, in a plethora of saddle-points, where a good grand strategy was a pure strategy. The principle is entirely general, and probably obvious: If a game has several solutions, each being a pure or a mixed strategy, then these in turn may be mixed, in any way you choose.

EXAMPLE 28. THE ADMINISTRATOR'S DILEMMA

The leader of the gang exhibited his openers and extended both hands to rake in the modest pot. He froze in the act as the new member—Cecil the Trigger—whipped something from his pocket and plunged his hand below the edge of the table.

"Well, Cecil?"

"I was just thinking, Paul. It would be nice for you to give me a present, let's say in honor of my joining the crowd—whatever swag (the gang is British) you happen to have on you, and that sparkler would be suitable. Don't waste your time wondering what's in my hand. It's just a ..."

Now it may be that Cecil is about to say 'gun.' On the other hand, he may name some innocuous object; perhaps a pipe. In either event, Paul is faced with a complex administrative decision. What are some of the possibilities?

Paul feels that the money is relatively unimportant: the important stakes are his life and prestige. Beginning with the grimmest possibilities, he lists:

(*a*) Facing up to a gun and getting shot.
(*b*) Backing down from a pipe which Cecil says is a pipe.
(*c*) Backing down from a pipe which Cecil says is a gun.
(*d*) Backing down from a gun which Cecil says is a gun.
(*e*) Backing down from a gun which Cecil says is a pipe.
(*f*) Facing up to a pipe which Cecil says is a pipe.
(*g*) Facing up to a pipe which Cecil says is a gun.

Paul decides to rate these on a scale running from 0 to 10, the horrible (*a*) being 0 and the triumph (*g*) being 10—no difficulty there. The assignment of values to the others is an unsatisfying art, which may make you unhappy; just remember that Paul is suffering too.* He decides to rate being completely duped at 5, and the intervening possibilities at 6, 7, 8, and 9. The strategies open to Cecil are four: he may flash a gun, or not; and he may say he has a gun, or not. Paul also has four: he may believe, or disbelieve, whatever Cecil says; or he may believe the statement 'gun' and disbelieve the other, or vice versa. He ends up with this game:

Cecil

		g,G 1	*g,P* 2	*p,G* 3	*p,P* 4
BG, BP	1	7	0	6	9
BG, DP	2	7	8	6	5
DG, BP	3	0	0	10	9
DG, DP	4	0	8	10	5

Paul (is to the left, spanning rows)

where

> *g* and *p* indicate gun and pipe
> *G* and *P* indicate Cecil's claims
> *B* and *D* indicate belief and disbelief

Thus Cecil's second pure strategy is to point a gun and claim it is a pipe. Paul's first pure strategy tells him to believe the statement—his last act, it turns out.

---

* There are two sections in Chapter 5, beginning on page 193, in which the importance of measurement, or estimation, accuracy is touched upon.

The solution runs along standard lines. We try to solve the 4 × 4 and our methods yield nothing but trouble; so we drop to a 3 × 3. Because Paul 3 and Cecil 3 appear somewhat unattractive, we try to do without them, and succeed. A solution for Paul is the mixture 7:17:0:4. The value of the game is 6. Cecil can hold Paul to this by playing 0:1:0:2. (Besides 7:17:0:4, Paul has these solutions: 1:3:0:0 and 3:21:4:0.)

It may seem astonishing that Paul doesn't stick to Paul 2, i.e., believe 'gun' and disbelieve 'pipe.' This is because he attaches value to prestige. If he did not rate prestige so high—in terms of our measurement scale, if he placed all the payoffs (except 0) near 10, but preserved the order—then the odds favoring Paul 2 would increase. For example, if he rates prestige factors rather lightly, he may use

$$0, 9.5, 9.6, 9.7, 9.8, 9.9, 10$$

as payoffs (instead of 0, 5, 6, 7, 8, 9, 10). A solution in this case turns out to be 291:9411:0:192—practically 3:94:0:2—and another is 3:99:0:0; so he still should not follow Paul 2 blindly. In this game, Cecil should play 0:4:0:98; i.e., he should always say he has a pipe, and should usually tell the truth.

EXAMPLE 29. THE COLONEL BLOTTO PROBLEM

The Task Force Commander, who had feared he would not make contact with the Reds, now feared he had made too much contact; for his companies at both advance points had become fully engaged, within a 10-minute interval.

When he planned this operation, G-2 had estimated the enemy at 600 men, whereas his own Intelligence now fixed the number at 1000 (probably 5 companies). So his two battalions of riflemen (6 companies) appeared to have an edge of about a company, instead of the comfortable factor of two he had planned on. His orders were to "destroy the maximum number of the enemy, at the least cost to your Command." The General Staff observer, a colonel who had attached himself for the duration of the mission, had been quietly amused when the Commander read this impossible directive to his battalion and company commanders, which served to confirm the Commander's views on visiting General Staff personnel.

His Baker and Dog companies were engaged at some distances to the northeast and to the northwest, respectively, of his main body. The remaining four companies were at hand and could be used in support of either point. As he waited for his battalion commanders to join him, he wondered what the General Staff man—now scratching his back against the tree—would do if the responsibility were his. He would probably look less relaxed, by God.

"And how does it look to you, Colonel? Remind you of anything out of the Second Punic War?"

"Curiously enough, it does remind me of something—something more remote than the Second Punic. Ever hear of Game Theory?"

"The term is not even hauntingly familiar."

"I spent some time one summer with some people—by the way, you familiar with the Zombie?—well, among other things, there was

a lot of talk about this Game Theory, a mathematical business supposed to have some bearing on warfare. One of the illustrations was one called Colonel Blotto's Problem—you'll pardon me; fairly typical of their irreverent attitude, I'm afraid; very undisciplined crowd. Anyway, the curious thing is that Blotto was faced with much this same situation."

"What did the Theory suggest he do? Negotiate?"

"Worse than that. They say—remember, this isn't *my* idea—they say you should keep your eye on that ant—no, the one on your map case. When it reaches the grease spot, look at the second hand on your watch. If it points to 6 seconds or less, you should divide your force equally between the threatened points. If it reads between 6 and 30 seconds, give the entire reserve to Dana; if between 30 and 54 seconds, give it to Harry; if between 54 and 60, pick out a new ant."

"Now isn't that a hell of a thing? . . . Well, let's go."

The analysis runs like this: There are 6 Blue and 5 Red companies. They are engaged at two points, in company strength. The alternatives assumed to be open to Blue are to divide his force in any of these ways: 5:1, 4:2, 3:3, 2:4, or 1:5. Red may divide his 4:1, 3:2, 2:3, or 1:4. We assume that equal forces result in a draw and that a superior force overwhelms the foe. The payoff is taken to be the number of companies overwhelmed, minus the number lost; this seems to meet reasonably well the spirit of the occasion, and it cannot conflict with the orders of Blue, which are strictly meaningless. The payoff matrix is:

Red

	41	32	23	14
51	4	2	1	0
42	1	3	0	−1
Blue 33	−2	2	2	−2
24	−1	0	3	1
15	0	1	2	4

(We have used 51 to denote the 5:1 division of forces, etc.)

Finding the solution is a little troublesome. However, eventually we get around to trying the subgame

Red

	41	32	23	14
51	4	2	1	0
42	1	3	0	−1
Blue 33	−2	2	2	−2
15	0	1	2	4

which is obtained by dropping the fourth Blue strategy. The all-strategies-active method yields a solution which satisfies the original game, namely 4:0:1:0:4 for Blue and 3:48:32:7 for Red.

From the symmetry of the game it is evident that another, but similar, solution would turn up if we dropped the second Blue strategy, rather than the fourth. In fact, this leads, as before, to 4:0:1:0:4 for Blue, but the solution for Red is now 7:32:48:3.

While there is no great virtue in doing so—unless it is to make the odds easier to look at—you may combine* these two Red solutions into 10:80:80:10, i.e., into 1:8:8:1. Then the results may be exhibited as follows:

		Red				Blue
		41	32	23	14	Odds
	51	4	2	1	0	4
	42	1	3	0	−1	0
Blue	33	−2	2	2	−2	1
	24	−1	0	3	1	0
	15	0	1	2	4	4
Red Odds		1	8	8	1	

Thus Blue should favor the extreme division of his force, over the equal division, with odds of 8:1, and he should never use the intermediate division. Red on the other hand should favor the more nearly equal division over the extreme division.

The value of the game to Blue is $1\frac{4}{9}$—between one and two companies. The business about the ant, the map case, and the second hand was a device for producing odds of 4:1:4, those needed by Blue.

The example may be carried another step. Suppose the natural sequel to the two battles is a third battle, between the survivors. We may then ask: How should the commanders plan the initial battles so as to maximize their expectations at the end?

According to our ground rules, the side which enters the final

* See page 191 on Multiple Solutions.

battle with the greater number of companies will overwhelm the other side; so its final payoff will be equal to the original strength of the enemy minus any friendly companies lost in the first two engagements. If the survivors are equal, no further company losses will occur. From these rules and the original matrix, it is easy to arrive at one which characterizes the new situation:

		Red			Blue
	41	32	23	14	Odds
51	5	4	4	4	10
42	5	5	3	−1	0
Blue 33	−5	5	5	−5	1
24	−1	3	5	5	0
15	4	4	4	5	10
Red Odds	2	19	19	2	

Here again the solution for Red is obtained by combining two solutions, namely, 1:7:12:1 and 1:12:7:1. The value is $8\frac{5}{21}$.

Comparing these results with those for the two-engagement battle, we see that the over-all battle requires that Blue pay greater heed to concentrating his forces against one point; the odds favoring concentration are now 20:1, instead of 8:1.

EXAMPLE 30. MORRA

There is a game called Morra, which, since it has a name, we did not invent. It will illustrate both a strength and weakness of Game Theory: The strength will be that the Game Theory strategy is needed by the players, for they are not likely to hit on a good scheme of play by chance; and the weakness is that a good strategy is formidable to compute.

The game is this: Each player extends some fingers and, simultaneously, guesses how many the enemy is extending. The number he may extend is 1, 2, or 3. If only *one* player guesses the enemy digits successfully, the payoff to him is the *total* number of digits extended on the play. Otherwise, the payoff is zero. (Payments are frequently in coin, rather than fingers.) Thus if Blue holds out 3 fingers and guesses 1, while Red holds out 1 and guesses 2, Blue will win because he's right; he will win 4 units of payoff.

Each player may extend 1, 2, or 3 fingers and may guess, reasonably, that the enemy is extending 1, 2, or 3. So there are nine possibilities open to him, i.e., nine pure strategies. We will identify these by a two-digit number such as 32, where the first digit represents the number he extends and the second represents his guess. Labeling the strategies in this way and calculating the payoffs, we represent Morra by the following 9 × 9 matrix:

Red

	11	12	13	21	22	23	31	32	33
11	0	2	2	−3	0	0	−4	0	0
12	−2	0	0	0	3	3	−4	0	0
13	−2	0	0	−3	0	0	0	4	4
21	3	0	3	0	−4	0	0	−5	0
22	0	−3	0	4	0	4	0	−5	0
23	0	−3	0	0	−4	0	5	0	5
31	4	4	0	0	0	−5	0	0	−6
32	0	0	−4	5	5	0	0	0	−6
33	0	0	−4	0	0	−5	6	6	0

Blue (label for rows)

A flash of genius is a useful thing at this point, because straight calculation is wretched. Our methods yield a complete set of zero odds when applied to the 9 × 9—apparently nobody wants to play. This means that one or more of the pure strategies must be excised, after which the thus-reduced game must be solved. There are 9 8 × 8's to consider (when we take advantage of symmetry, so there won't be 81), 36 7 × 7's, and 84 6 × 6's. We didn't do it that way.

For reasons bordering on numerology, and mildly aided by a knowledge of the answer, we became interested in the 126 5×5 games which are symmetric in the strategies. We were lucky enough to find this one—from which the first, fourth, sixth, and ninth strategies have been excised—at the third trial: Red

		12	13	22	31	32
	12	0	0	3	−4	0
	13	0	0	0	0	4
Blue	22	−3	0	0	0	−5
	31	4	0	0	0	0
	32	0	−4	5	0	0

Using our all-strategies-active method, we grind out these odds (the common factors have been eliminated):

$$0:5:4:3:0$$

We test and find that these do satisfy the 5×5 subgame and, moreover, the original 9×9. So a sound way to play Morra is

Strategy:	11	12	13	21	22	23	31	32	33
Odds:	0 :	0 :	5 :	0 :	4 :	0 :	3 :	0 :	0

The three noticeable component strategies are (1) to extend 1 finger and guess 3 for the other fellow, or (2) to extend 2 and guess 2, or (3) to extend 3 and guess 1; and to mix these three according to the odds 5:4:3. The value of the game is zero, of course, it being a fair game. It is probably a good game to teach to your friends, since the solution is easy to memorize, and yet difficult to intuit.

EXAMPLE 31. THE MAZE

A cat and a mouse enter this blind maze.

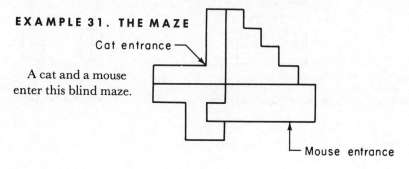

They may turn corners, but cannot turn around. Traveling at the same speeds, they are given enough time to cover one-fourth of the maze, after which the mouse is rescued, if he is available. What should their strategies be?

They of course instantly recognize that the maze is equivalent to this one:

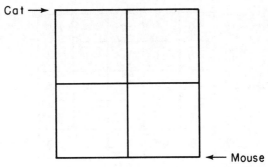

and therefore that the eight possible pure strategies look like this:

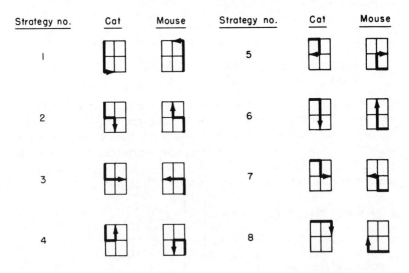

If we count cat-catches-mouse as 1 and cat-misses-mouse as 0, then the cat is the maximizing player and the mouse is the minimizing player. By comparing the various pure strategies, this payoff matrix is found:

Mouse

		1	2	3	4	5	6	7	8
	1	0	0	0	1	0	0	0	1
	2	0	1	1	1	1	1	1	0
	3	0	1	1	1	1	1	1	0
	4	1	1	1	1	1	1	1	0
Cat	5	0	1	1	1	1	1	1	1
	6	0	1	1	1	1	1	1	0
	7	0	1	1	1	1	1	1	0
	8	1	0	0	0	1	0	0	0

Now Cat 5 dominates Cat 1, 2, 3, 6, and 7; and 4 dominates 8. Mouse 1 is dominated by Mouse 5, and 8 by 4. Also, Mouse 2, 3, 6, and 7 are alike. Thus the game reduces to

Mouse

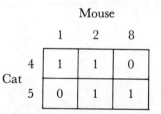

		1	2	8
	4	1	1	0
Cat	5	0	1	1

which is easy to solve: The cat should mix Cat 4 and 5 equally, and the mouse should mix 1 and 8 equally.

Referring back to the descriptions of the pure strategies, we see that the cat should travel along small loops, and that the mouse should stick to the outside of the maze.

EXAMPLE 32. MERLIN

"I need some help," Guinevere began. "That hussy Yseult has an uncanny knack of having her kerchief worn by the winner when the boys do a little jousting, and my champion is often horizontal at the end. It's practically insubordination, and it must stop."

Merlin's left hand became a toad; it croaked at him. He started slightly and sat up straighter.

"Pardon me, my dear, my mind must have wandered for a moment . . . Yes, the jousts. I haven't been following them lately. What are the present rankings?"

"It's terribly confusing," wailed Guinevere. "There have been only seven Senior knights in town lately, and most of their tilting has been indecisive. However, some definitely have a hex on others. You can rely on Lancelot to beat Tristram; Tristram to beat Gareth and Lamorak; Lamorak to beat Gawain and (so help me!) Lancelot; Gareth to beat Tor; Tor to beat Sir Kay and Tristram—I wonder if he's *really* so honest. Then Kay beats Gawain, and Gawain beats Gareth. It's maddening; they're so consistent in some ways, but . . ."

"But they're not transitive," murmured Merlin, as he took notes, which looked like this:

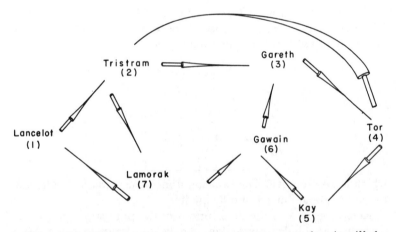

"That Tor-Tristram development isn't very neat, but it will do. Gwen, just make yourself comfortable for a few minutes—a subject of this difficulty deserves a trance."

Merlin slumped down with closed eyes. Lady Guinevere sat perfectly still, transfixed by the beady eyes of his familiar, the small owl who made its nest in his hair.

The old man shuddered, and sat up. "My," he sighed, "I always find it grueling when I have to visit—even briefly—the twentieth century; but I found what I wanted. There is a thing called—well, let's say Jousting Theory. I won't bother you with the details. Here's what you do: Put some stones, marked with the knights' colors, in an urn; also—and this is vital—a pinch of the finest unicorn horn, finely ground. Don't put in stones for Kay and Gawain. When you want to choose a knight, shake the urn, do two somersaults—one backward— and draw a stone. You will at least hold your own with Yseult."

"Why, that's wonderful, Merlin! How can I possibly repay you? Name your boon!"

"Well," said Merlin, "you know how it is: by and large I can arrange for most of my wants"—as he spoke he raised to his lips a slice of smoked salmon which had materialized in his hand. As he nibbled at it he went on: "Of course, you could do a somersault for me, as you go through the doorway. I'd like to see it, without having to arrange it."

Counting indecisive combats (draws) as 0 and wins and losses for Guinevere as 1 and -1, the payoff matrix is:

Yseult

		1	2	3	4	5	6	7
	1	0	1	0	0	0	0	-1
	2	-1	0	1	-1	0	0	1
	3	0	-1	0	1	0	-1	0
Guinevere	4	0	1	-1	0	1	0	0
	5	0	0	0	-1	0	1	0
	6	0	0	1	0	-1	0	-1
	7	1	-1	0	0	0	1	0

Here the strategy index identifies the knights.

The analysis is straightforward, though tedious. By exhaustively examining the 7 × 7 game, and the 6 × 6's contained in it—or by a little trial and error—you eventually get down to a 5 × 5 based on knights 1, 2, 3, 4, and 7. The solution to the 5 × 5 turns out to be to mix these five knights equally; and this also solves the original game. So both girls should use this mixed strategy: elect knights according to the odds 1:1:1:1:0:0:1.

SUMMARY OF 4 × m METHODS

Let us pause now to review our position on 4 × 4, and larger, games. Our procedure is to search, first, for a saddle-point; if one is present, we know that a satisfactory grand strategy is to use the pure strategy identified with the saddle-point. If the game does not possess a saddle-point, we clean out of the matrix all of Blue's dominated strategies and all of Red's dominating ones; unfortunately, our test for dominance is purely inspection, which is fallible in all cases—especially in those we have called cases of hidden dominance, where a *combination* of pure strategies dominates (or is dominated by) a pure strategy. In any event, having cleaned the matrix to the best of our ability, we then make the assumption, tentatively, that all strategies are active and we try to compute the best mix, using a very explicit and firm procedure. There is a reasonable prospect—perhaps a little more—that we will find ourselves in trouble when we finish, thanks to the number of pitfalls that beset the path.

What guise will this trouble wear? Actually, it has a varied wardrobe and many sets of false whiskers. We met it earlier, in a 3 × 3 game, when the oddments for all of Blue's strategies seemed to be zero, a patently false answer; the answer *may* be correct if the odds for some strategies are zero, but not when all of them are. A useful check (before common factors are divided out) is that the sum of the oddments for all of Blue's strategies must be the same as Red's total. If they are not, either the method has failed or the arithmetic is at fault; since you won't know which it is, the latter must be rechecked carefully. Again, trouble may first be noticed when you calculate the average payoff of the grand strategy against each pure strategy of the enemy—someone may be winning or losing more than he is entitled to, assuming proper play on both sides.

In all cases where trouble develops, once you are convinced the arithmetic is sound, the cure is the same—and unpleasant: Drop some strategies, solve the thus-reduced game, and try this solution in the original game. If the game is very large, you should venture further carefully, calling up all reserves of insight and luck that can be mustered, or quietly retire from the field; for the situation may be desperately out of control—there being an awful lot of subgames awaiting inspection.

If, in the original large game or in the one that remains after the matrix has been cleaned (by dominance arguments), one player has more pure strategies than the other, nothing new is required. Both players are bound to have good grand strategies which use no more pure strategies than are available to the player who has the fewer. It follows then that, to perform the all-strategies-active calculation, you must eliminate enough strategies from the repertoire of the player who has an excessive number to bring the game to square form. Thus a 5×10 game must be reduced to 5×5, 4×4, or even to a lesser game before you can hope to calculate the solution. Since there are many possibilities, the process may be tedious, perhaps prohibitively so.

With the heartening remarks of the last two paragraphs as sustenance, fortified with good will and Godspeed, we now offer to the brave among you some Exercises.

EXERCISES 5

Determine the oddments and values of the following games:

1.
Red

		1	2	3	4
	1	7	6	0	2
	2	3	8	2	5
Blue	3	1	0	2	2
	4	3	0	1	3

2.
Red

		1	2	3	4
	1	1	2	3	4
	2	8	7	6	5
Blue	3	7	6	5	4
	4	0	1	2	3

3.
Red

		1	2	3	4
	1	3	5	4	3
	2	3	4	2	1
Blue	3	1	1	3	4
	4	0	5	2	2

4.

		Red 1	2	3	4	5
Blue	1	2	2	0	1	0
	2	1	0	−1	1	−2
	3	1	3	−1	4	−1
	4	4	2	0	1	0

5.

		Red 1	2	3	4
Blue	1	2	3	1	4
	2	1	2	5	4
	3	2	3	4	1
	4	4	2	2	2

6.

		Red 1	2	3	4
Blue	1	1	7	0	3
	2	0	0	3	5
	3	1	2	3	0
	4	6	0	2	0

7.

Red

Blue	1	2	3	4
1	8	0	6	7
2	3	3	4	8
3	1	5	9	2
4	3	6	3	1

8.

Red

Blue	1	2	3	4
1	3	4	0	3
2	5	0	2	5
3	7	3	9	5
4	4	6	8	7

9.

Red

Blue	1	2	3	4
1	6	3	0	0
2	0	0	5	1
3	1	1	2	0
4	1	1	7	3
5	1	1	0	7
6	0	0	2	0

10.

	Red 1	2	3	4	5	6	7	8
Blue 1	1	1	1	0	1	6	0	0
2	7	2	0	0	2	0	1	5
3	0	3	2	3	3	2	3	3
4	3	0	8	5	0	0	4	1

11.

	Red 1	2	3	4	5
Blue 1	0	1	0	2	0
2	3	0	0	1	2
3	0	0	2	1	2
4	1	3	0	0	1
5	1	2	3	1	0

12.

	Red 1	2	3	4	5
Blue 1	3	3	4	4	8
2	8	0	6	8	7
3	1	5	9	5	2
4	3	6	3	1	1
5	5	5	0	3	0

13.

Red

		1	2	3	4	5
	1	0	4	8	4	1
	2	2	5	2	0	0
Blue	3	1	4	0	8	4
	4	7	−1	5	7	6
	5	2	2	3	3	7

14.

Red

		1	2	3	4	5
	1	2	7	7	6	7
	2	6	6	6	9	2
Blue	3	9	4	5	5	9
	4	4	9	3	8	6
	5	7	7	9	3	6

15.

Red

		1	2	3	4	5
	1	3	4	0	3	0
	2	5	0	2	5	9
Blue	3	7	3	9	5	9
	4	4	6	8	7	4
	5	6	0	8	8	3

16.

Red

		1	2	3	4	5
	1	2	2	5	1	3
	2	0	0	4	3	4
Blue	3	1	2	1	0	6
	4	5	4	5	0	5
	5	3	1	1	6	6

17.

Red

		1	2	3	4	5
	1	5	5	4	5	2
	2	5	4	7	1	6
Blue	3	2	3	4	1	7
	4	3	6	7	3	2
	5	4	4	3	0	2

18.

Red

		1	2	3	4	5
	1	6	4	9	3	7
	2	0	2	9	8	5
Blue	3	5	3	4	2	4
	4	1	6	2	1	8
	5	1	6	1	3	6

19.

Red

		1	2	3	4	5	6
	1	6	4	−2	4	0	5
	2	1	4	1	2	4	6
	3	1	1	4	1	2	−1
Blue	4	0	0	3	−1	1	3
	5	3	2	3	−2	3	−2
	6	1	−1	−1	4	4	−2

20.

Red

		1	2	3	4	5
	1	−2	3	1	−1	−1
	2	−1	−3	1	3	0
Blue	3	0	3	−3	−1	1
	4	3	−3	1	−1	0

21.

Red

		1	2	3	4	5	6	7
	1	8	3	1	3	2	5	3
	2	6	6	0	0	2	4	1
	3	0	3	5	6	6	7	6
Blue	4	6	4	9	3	1	0	6
	5	1	6	4	4	3	5	6
	6	7	8	2	1	5	0	0

22.

Red

		1	2	3	4	5
	1	3	1	4	1	6
	2	6	3	1	4	1
Blue	3	1	6	3	1	4
	4	4	1	6	3	1
	5	1	4	1	6	3

CHAPTER 5

Miscellany

This chapter has the characteristics of a catchall. It includes game-solving devices and aids, as well as brief mention of subjects the full discussion of which would take us well beyond the proper range of the book. Some of the items included have already put in a specter-like appearance, usually at times when it was not expedient to explore each dimly lit corner and thus lay the ghosts.

We begin, however, with a subject which is a somewhat logical continuation of the last chapter, though it represents a complete break in method.

APPROXIMATIONS

We have grubbed along now for several chapters, working into larger and larger games; and we have reached the point where we can assert that we know how to go about solving an arbitrarily large one. Perhaps more accurately, we have methods which, when applied to a game whose strategies and payoffs are listed, will yield a solution. This is impressive enough to warrant mutual congratulations.

Having said that, we must now go on and admit that general conditions are deplorable. Our position is somewhat comparable to that of a clerk who undertakes to satisfy the National Debt by disbursing one-dollar bills: the unit operation, counting out bills at, say, one per second, is feasible and well understood; but our clerk has—what with vacation, sick leave, and the forty-hour week—a forty-thousand-year task. In our case, when the number of strategies available to each player is somewhere between four and ten—exactly where the break comes depends on the power of your intuition and on the quality of your luck—the industry required becomes prodigious. For instance, the number of 4×4 subgames in a 10×10 game exceeds forty thousand.

In view of this difficulty with games which lie on the sunny side of 10×10, it may occur to you that even the professional game solver will blanch slightly at the task of getting exact solutions to games in the range from 10×10 to 100×100. As a matter of fact, his pallor is awful. To illustrate how bad things can be: The number of subgames which may have to be examined in a 130×130 game is something like a one with seventy-eight zeros after it, which is truly a magnificent number; the proverbial astronomical number at last, for it coincides with Eddington's estimate of the number of protons in the universe.

It is evident that we are, in some way, requiring too much. A scientific theory should do more than provide us with impossible tasks, and examination of all the subgames present in a large game is strictly impossible. So we need either a more powerful method of analysis or a less demanding requirement; and, being mathematical cowards, we prefer the latter.* A great amelioration of our situation will come about if we back off on the requirement for an exact solution and content ourselves with an approximate one.

The need to make such a compromise is not novel, nor do we thereby lose any practical utilities. The situation is comparable to that faced by a man who is given a long board, a saw, a measuring tape, and orders to cut five lengths which can be formed into a square having a diagonal brace—the parts to fit exactly. If he is a mathematician, the poor fellow just can't do it, of course, because the brace must be longer than the sides by a factor equal to the square root of

*Ten years later, we set forth the more powerful method in Chapter 6. However, the approximate method is still of interest, depending on the magnitude of the game and on the kind of computing equipment available to the game solver.

2; and this number is not marked on any tape. However, laying aside his mathematical toga and the requirement that they fit exactly, he can provide parts which will satisfy any reasonable demand.

In approaching the subject of approximations, we must make a preliminary decision. Are we interested in finding a mixed strategy which resembles, physically, the best one? Not necessarily; while good form is frequently an admirable quality, as in cricket, our chief interest may be only that the value of the game be attained—approximately. This attitude will give us some latitude, for there may be many strategies which yield a reasonably good payoff; and it would only reduce that number if we were to require that they be beautiful, too. And the more there are, the easier it ought to be to find one.

We shall describe a method which is both powerful and mathematically respectable. You may think of it as a succession of plays of a game; and at each stage the players make the countermove which looks best in view of what the enemy has done to date. It is most easily exhibited by using a specific matrix, but it is completely general and will work for any matrix. Take, then, the game:

Red

		1	2	3	4
	1	2	3	1	4
	2	1	2	5	4
Blue	3	2	3	4	1
	4	4	2	2	2

Mark any *row* you wish (the first, say) by putting an asterisk at the end of it. Then copy that row below the matrix, and mark the *smallest* number in it:

2	3	1	4	*
1	2	5	4	
2	3	4	1	
4	2	2	2	

2	3	1*	4

This completes the first step in the approximation process. Continuing, the number 1* marks the third *column;* copy that column at the right, and mark its *largest* number, 5*:

2	3	1	4	*	1
1	2	5	4		5*
2	3	4	1		4
4	2	2	2		2
2	3	1*	4		

This 5* marks the second *row; add* that row, piece by piece, to the row already below the matrix, and mark the *smallest* number, 3*, in the new row:

2	3	1	4	*	1
1	2	5	4		5*
2	3	4	1		4
4	2	2	2		2
2	3	1*	4		
3*	5	6	8		

This completes the second step in the process and we are solidly in business: just keep going. The next step, since 3* marks the first column, is to add that column, piecewise, to the column already at the right:

2	3	1	4	*	1	3
1	2	5	4		5*	6*
2	3	4	1		4	6
4	2	2	2		2	6
2	3	1*	4			
3*	5	6	8			

We mark its greatest number, 6*. (Here several numbers tie for greatest; it doesn't matter which is marked.) Since 6* marks the second row, we add that row to the bottom row, completing the third step, and so on. We carry out the process for ten steps to give a better view of it. (Note that we do not place a mark in the last column, the last numbers written.)

2	3	1	4
1	2	5	4
2	3	4	1
4	2	2	2

```
* 1   3   5   7   10  13  16  19   22   25
 5*  6*  7   8   10  12  14  16   18   20
 4   6   8  10   13  16  19  22*  25*  28
 2   6  10* 14*  16* 18* 20* 22   24   26
```

```
 2    3   1*   4
 3*   5    6    8
 4*   7   11   12
 8*   9   13   14
12   11*  15   16
16   13*  17   18
20   15*  19   20
24   17*  21   22
26   20*  25   23
28   23*  29   24
```

A reasonable question at this point is: So what? We have a smooth mechanical process, which could be subcontracted to child labor, and no clue to its significance. Let us continue, in this detached fashion, just one step further: Count the asterisks after each row and column, and write the totals next to the matrix:

		Red			Blue Odds
	1	2	3	4	
Blue 1	2	3	1	4	1
2	1	2	5	4	2
3	2	3	4	1	2
4	4	2	2	2	5
Red Odds	3	6	1	0	

These totals are *estimates* of the odds according to which Blue and Red should mix their strategies; in this case, 1:2:2:5 and 3:6:1:0, respectively.

How good are these estimates? We can judge this by observing how bad things can be for players who use these mixed strategies. Referring to the calculation, we see (from the bottom row) that Blue's winnings in 10 plays may be as small as 23; this would occur if Red consistently used his second strategy. So Blue's average winnings for 10 plays may be as small as 2.3. Similarly, from the column at the extreme right, we see that Red's mix could cost him (after 10 plays) as much as 28 (if Blue played Blue 3 every time), an average of 2.8 per play.

The spread between 2.3 and 2.8 reflects imperfections in the strategy mixes used; for if the grand strategies were optimum, each player would guarantee himself average winnings (or losses) equal to a number which lies between 2.3 and 2.8.

The range of uncertainty, 2.3 to 2.8, is about 20 per cent of the payoff, which is rather large. Better strategies can be found by continuing the estimation process. We find, after building the total frequencies up from 10 to 20 (as indicated here),

2	3	1	4	25	29*	32*	35*	38*	39	41	43	45	49	53
1	2	5	4	20	24	26	28	30	35	36	37	38	42	46
2	3	4	1	28*	29	32	35	38	42*	44*	46*	48*	49	50
4	2	2	2	26	28	30	32	34	36	40	44	48	50*	52

28	23	29	24
30	26	33	25*
32	29*	34	29
34	32*	35	33
36	35*	36	37
38	38	37*	41
40*	41	41	42
42*	44	45	43
44*	47	49	44
46	50	53	45*
50	52	55	47*

that the strategies for Blue and Red have changed to 5:2:7:6 and 6:9:2:3, respectively, and the average winnings and losses are now confined to the range from $4\frac{7}{20}$ to $5\frac{3}{20}$, i.e., from 2.35 to 2.65. So the uncertainty in the payoff has been reduced from some 20 per cent to about 12 per cent.

This may be satisfactory; if not, the process may be continued indefinitely. As the frequencies increase, fluctuations in the guaranteed payoffs will occur; but in general the strategies found will yield payoffs certified to approach, more and more closely, the value of the game.

The exact solution to this game is 8:3:7:9 for Blue and 5:7:3:3 for Red; and the value to Blue is $2\frac{5}{9}$.

When the payoff matrix is built of small integers, this method permits us to face 10 × 10 games, and even larger ones, with equanimity; for only modest quantities of paper, pencil, and time are needed.

MORE ON DOMINANCE

The discussion of dominance in earlier chapters was quite sketchy, so as not to disturb the general line of thought; that danger being past, we may now fill in the blanks. To avoid tiring repetitions from a slightly different viewpoint, we shall phrase all items in terms of Blue's problem and trust that you will make the necessary modifications when applying them to Red's.

We have learned to drop instantly any strategy which is dominated by another; thus, if the payoffs in one row are smaller than the *corresponding* ones in another, we omit the former row. You may do this even if some of the corresponding payoffs are equal. When all are smaller, the situation is called one of *strict* dominance; when some are equal, it is called (of course!) *nonstrict* dominance. The reason it is worth while to distinguish these by inventing special names is this: Eliminating a row by nonstrict dominance may make a subtle change in the game, in that the new game may have fewer solutions than the old one. Since we have devoted our attention exclusively to finding one solution, rather than to finding all solutions, we have not needed to discriminate the kinds of dominance.

It has also been mentioned that a row may be dominated by a felicitous combination of rows; thus in

Red

		1	2	3	4	5
	1	2	3	7	0	2
Blue	2	3	5	7	1	1
	3	2	2	7	2	4

two parts (i.e., by weight) of Blue 1 are dominated by one part of Blue 2 mixed with one part of Blue 3.

Further, if some row dominates a mixture of others that uses the same total number of parts, or odds, then a row in the mixture is unnecessary.

There is no rule of thumb for identifying these situations, and they are not always obvious; indeed, it isn't reasonable to expect to have a simple rule, for the problem is similar to that of finding a good mixed strategy for the game. However, a little looking is worth while. Thus in Example 16, The Heir (page 101), row 1 is readily seen to be dominated by a ½:½ combination of rows 2 and 3, and should therefore be omitted.

We mention one final type of dominance—the matrix type—which is similar in spirit to the saddle-point idea; in fact it is a generalization of the saddle-point. Let us consider, briefly, the following unfriendly-appearing object:

1	3	1	4	6
2	0	3	3	1
3	1	3	5	4
−1	−1	6	−1	2
0	−4	0	3	4
2	−1	10	1	2
−3	0	3	4	5

There is a very good reason for subdividing it into four components, as follows:

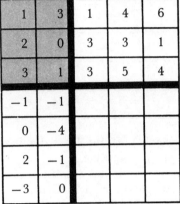

The shaded matrix now has this important property: *each* column of the matrix on its right dominates *some* column of the shaded matrix; and *each* row of the matrix below is dominated by *some* row of the shaded matrix. The numbers of the remaining matrix are unimportant when the shaded matrix has the properties just stated; it doesn't matter at all what they are, so we do not bother to write them. As a matter of fact, when the shaded matrix has the properties noted, the precise value of all the other numbers becomes unimportant and the entire game reduces to the shaded-matrix game:

1	3
2	0
3	1

This, obviously, was worth looking for. In searching for things like it, remember that you may freely interchange any rows, or any columns. While this trick is worth knowing, you will not usually achieve a happy ending lightly; in fact, you may have to forsake your usual haunts and just live with a big matrix, before it will yield—if it will yield.

SIMPLE SOLUTIONS

We have usually abstained from stating rules which apply only to very special games. However, the introduction of one or two at this late stage should not cause confusion.

A game is said to have a *simple* solution if either player may play, with impunity, *every* pure strategy against the enemy's good grand strategies. A particular case of this is the game in which the players' grand strategies contain all their pure strategies; of course we have used the impunity principle to test a proposed grand strategy for excellence. The simple-solution idea is therefore more comprehensive than the all-strategies-active idea, for it includes some games having pure strategies that are 'played' with zero odds. Thus, in a 4×4 game, for example, Blue may play any pure strategy against Red's grand strategy if their respective grand strategies are 3:2:1:7 and 1:5:2:4; he may do the same when the grand strategies are 3:0:1:7 and 1:5:0:0, *provided* the game has a simple solution (and this is it).

For example, this game has a simple solution, because each player may use *any* pure strategy against the other player's good mixed strategy:

	Red 1	Red 2	Red 3	Blue Odds
Blue 1	6	− 10	3	0
Blue 2	4	4	4	1
Blue 3	2	11	5	0
Red Odds	1	0	2	

The average payoff will always be 4.

As usual, we have no sure-fire method of recognizing, in advance, that a game has a simple solution; but there are some cases where recognition is possible. One is this:

The sum of the payoffs in rows is the same for all rows, *and* the sum in columns is the same for all columns. Thus in

	Red 1	Red 2	Red 3	Red 4
Blue 1	1	4	2	1
Blue 2	− 2	4	2	4
Blue 3	7	− 2	2	1

the row sums are all 8 and the column sums are all 6. In games such as this, there is a simple solution and it is easy to find it; in fact, it is an all-strategies-active solution, with equal odds; thus Blue 1:1:1 and Red 1:1:1:1 form a solution. The value of the game is found by dividing a column sum by the number of Blue strategies or a row sum by the number of Red strategies, e.g., ⁶⁄₃ or ⁸⁄₄ here.

There is one other class of simple-solution games which we can recognize, but which will require a *tour de force* by writer and reader. We must learn the meaning of *diagonal* and *separated*.

The diagonals of a matrix, say of

3	0	2	8
9	2	3	0
0	7	2	3
2	3	6	2

may be identified by writing the matrix a second time, under the first, and by then drawing sloping lines, so:

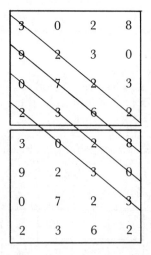

(Once you understand how it goes, it really isn't necessary to duplicate the matrix.)

A diagonal is said to be *separated* from the remainder of the matrix if it is made up exclusively of large numbers: its smallest element must be at least as large as the largest in the remainder of the matrix.

We shall also need a magical number, say M, obtained by multiplying together the number of rows in the matrix and the largest number in the remainder of the matrix (the part left after omitting the separated diagonal). Now we are ready for our class of simple-solution games.

If the matrix contains a separated diagonal, and if the sum of the payoffs in each row, *or* in each column, is at least as large as the magic number M, then—so help us—the game has a simple solution.

In the present case, the 9-7-6-8 diagonal is separated. The number of rows is 4, and the largest payoff remaining is 3; so $M = 4 \times 3 = 12$. The row and column sums are 13, 14, 12, 13 and 14, 12, 13, 13—all (rows *and* columns, which is more than we need) at least as large as $M = 12$. Thus the game has a simple solution. Easy, isn't it?

MULTIPLE SOLUTIONS

While we have devoted our attention exclusively to the problem of finding *some* solution to each game, it has been evident at times that the games may have more than one solution.

The actual situation is, in a sense, explosive: Every game has either exactly one solution or infinitely many. We haven't undertaken to prove many statements in this book, but the last one is so easy as to be irresistible: Assume the game has at least two solutions (i.e., grand strategies); then these may, in turn, be mixed according to any odds —and there are infinitely many possible odds.

It may seem from this that counting solutions is not a very satisfying task—as in testing gunpowder, the activity is either minor or excessive. However, there is a way of looking at it which makes it more meaningful: Each game has only a finite number of *basic* solutions.

Basic solutions are defined in this manner: If a square submatrix (i.e., subgame) of the game has a unique simple solution—i.e., one such that each of the other player's pure strategies may be used against it without unnecessary loss—and if that simple solution also satisfies the original game, then it is called a basic solution. A game has only a finite number of square submatrixes (not all of which will have simple solutions); so it must have only a finite number of basic solutions. The totality of solutions is then obtained by making all possible mixtures of these basic solutions.

Here is an example:

Red

		1	2	3
	1	2	6	0
Blue	2	5	3	6
	3	4	4	3

This has no saddle-points (which, incidentally, would be basic solutions). Of the nine 2 × 2 submatrixes, these two

2	6
5	3

6	0
3	6

contain completely mixed (and therefore simple) solutions, and ones which satisfy the original game; they are Blue 2:4:0 vs. Red 3:3:0 and Blue 3:6:0 vs. Red 0:6:3. (The common factors have been left in for ease of verification.) The 3 × 3 matrix also possesses a simple solution, and since this is itself the original matrix, the simple solution is basic; however, it happens to be identical with one of the earlier solutions (the first one). So the game has only two distinct basic solutions which, when the common factors are eliminated from the odds, may be written as

$$\text{Blue 1:2:0 vs. Red 1:1:0}$$

and

$$\text{Blue 1:2:0 vs. Red 0:2:1}$$

So Blue has a unique grand strategy, namely, 1:2:0, whereas Red may use 1:1:0 or 0:2:1, *or any mixture of these.* For example, by assigning a weight of 3, say, to his first basic strategy and of 5, say, to his second, he gets 3:3:0 and 0:10:5, which combine to give Red 3:13:5, a perfectly good grand strategy for his use.

The confirmed game addict will not be able to resist searching for the basic solutions to the following game. Blue has five and Red has one; they are given in the back of the book.

EXERCISE 6

Red

		1	2	3	4
	1	16	−8	9	−3
	2	−20	4	9	−3
Blue	3	25	1	18	−6
	4	−11	13	−18	6

ON MEASUREMENT

An obvious difficulty, which the user of Game Theory must learn to live with, has to do with the need to measure things which do not afford secure bases for measurement. This is a serious difficulty, and one which must not be underestimated. On the other hand, it is almost as easy to overestimate it, if you do not appreciate the process by which the measurements determine the strategies and the outcome of the game. Let us review, and extend somewhat, that which is relevant to this subject.

It is evident from the discussions of dominance that situations may arise in which precise measurement is not everywhere required. It may be possible without actual measurement (e.g., by qualitative comparison) to establish that some row, or column, or submatrix is dominant, and the elements thus eliminated may have been difficult to measure; after all, it is not necessary actually to measure all trees in a forest in order to find the height of the tallest.

We observed, early in our study, that a player's grand strategy is unaffected when a constant is *added* to each payoff of the game matrix. Therefore, in particular, if some arrangement of four distinct numbers—say 1, 7, 11, and 1066—comprises the matrix, we shall not misplay the game if we misestimate the numbers by 533—and use −532, −526, −522, and 533. We also observed that the play is unaffected when the numbers in a matrix are *multiplied* by a positive constant. In particular, the last set could be multiplied by ⅓, giving −40¹²⁄₁₃, −40⁶⁄₁₃, −40²⁄₁₃, 41. This process of adding constants and multiplying constants can be continued, in any order. Thus we are free to add 40 to the last sets, and so we know that the games

1	7	1
11	1	1066
1066	1	7

and

$-12\frac{2}{13}$	$-6\frac{9}{13}$	$-12\frac{2}{13}$
$-\frac{2}{13}$	$-12\frac{2}{13}$	81
81	$-12\frac{2}{13}$	$-6\frac{9}{13}$

say, are the same from the point of view of play.

It is clear, then, that the invariance of game strategies with respect to addition and multiplication of constants ameliorates, to some extent, the cross of measurement accuracy. The exact extent may be appreciated by considering a payoff matrix which contains three distinct values: it will not matter what number is assigned to the least valuable, nor to the most valuable; all that matters is the *relative* position of the third one.

In passing: There is one aspect important to planning which *is* affected by the things which do not affect the strategies, namely, the value of the game. It may be significant to know—in fact it may affect one's intention to engage in the game, provided one is free to choose—that the game is worth 20 instead of 10, or 10 instead of 20; but the same good strategy will win whichever quantity is available.

So the place where game strategies may be sensitive to errors of measurement may be thought of as the middle ground, that region between the extreme values; for the positions of the extremes may be shifted at will without effect, so long as the others are shifted appropriately, too. To gain some experience with the interplay of measurement errors, strategies, and value of the game, we give the results of a small-scale experiment:

We begin with a typical 3 × 3 game,

Red

		1	2	3
	1	6	2	0
Blue	2	4	8	4
	3	3	3	10

The payoffs range from 0 to 10, and in what follows we shall constrain all games to that range; this eliminates certain scale effects, such as addition and multiplication by constants, from the value of the game. We now simulate measurement errors by changing, independently, each element in the matrix as follows: We add -1, 0, or $+1$, and decide which to do on the basis of equal odds; thus we are simulating a uniform random error. We make an exception to the rule: never change the original 0 or 10, and reject any changes which will force some payoff outside this range—the purpose here is to retain the original 0-to-10 scale. The random-number table we happened to consult transformed the above game into this one:

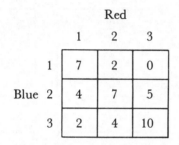

Red

		1	2	3
	1	7	2	0
Blue	2	4	7	5
	3	2	4	10

This process could evidently be continued: The last matrix could be subjected to a random shock, of the type described, and another matrix produced; this one would presumably be less strongly related to the original than was the second, and so on. Or the third one may be thought of as the true matrix, and the first one as being derived from it by two random shocks. We shall find it convenient, in what follows, to adopt the latter view.

Now, the question of interest is: Suppose we believe we are involved in one game, say the first, when in reality we are playing some other, say the second, or the third; how serious are the consequences of the misconception? We confine the investigation to Blue.

The second game (the true one) requires that Blue play his three pure strategies according to the odds 22:26:16, which will yield him $4\frac{17}{32}$, the value of the game; but Blue, who is guided by the error-ridden first matrix, believes he should use the grand strategy 7:7:6, and so he does. The consequences will depend on what Red does; if the latter uses Red 1, 2, or 3, Blue will win $4\frac{9}{20}$, $4\frac{7}{20}$, or $4\frac{15}{20}$, respectively. The worst that can happen is that he will win $4\frac{7}{20}$, which is 0.18 less than the $4\frac{17}{32}$ he could have won had he known the true matrix.

Our experiment was carried out in the above manner. The first matrix was progressively shocked through a sequence of fourteen forms:

1.

6	2	0
4	8	4
3	3	10

2.

7	2	0
4	7	5
2	4	10

3.

6	2	0
3	6	4
1	3	10

4.

7	1	0
4	7	3
0	4	10

5.

6	2	0
3	7	3
1	5	10

6.

5	3	0
3	7	4
1	4	10

7.

4	3	0
3	7	3
2	5	10

8.

3	3	0
2	8	4
2	4	10

9.

2	3	0
1	8	4
3	3	10

10.

1	2	0
1	8	4
3	4	10

11.

1	1	0
0	8	4
2	5	10

12.

1	2	0
0	8	3
1	4	10

13.

0	3	0
1	8	4
2	5	10

14.

0	2	0
1	7	3
1	5	10

15.

0	3	0
2	7	4
0	4	10

then each game, of the last ten, was treated as the true game corresponding to the five that preceded it.

The good strategies for these games are

Game	Blue	Red	Value
1	7:7:6	14:2:4	$4\frac{2}{3}$
2	22:26:16	40:5:19	$4\frac{17}{32}$
3	25:24:14	36:10:17	$3\frac{47}{63}$
4	34:32:27	51:9:33	$3\frac{29}{31}$
5	36:12:24	41:15:16	$3\frac{5}{8}$
6	33:3:18	36:4:14	$3\frac{3}{5}$
7	4:0:2	5:0:1	$3\frac{1}{3}$
8	8:0:3	10:0:1	$2\frac{8}{11}$
9	0:0:1	1:0:0	3
10	0:0:1	1:0:0	3
11	0:0:1	1:0:0	2
12	0:0:1	1:0:0	1
13	0:0:1	1:0:0	2
14	0:0:1	2:0:0	1
15	0:1:0	1:0:0	2

where common factors have been left in, so that the odds for Blue and Red have the same total.

With this information, the calculation is straightforward.

Average Amount of Payoff Lost by Blue

The game has been subjected to this number of random shocks

		1	2	3	4	5
	1	.18	.30	.09	.38	.46
	2	.21	.22	.30	.37	.24
	3	.57	.47	.21	.16	.33
	4	.32	.41	.26	.36	1.05
Blue's strategy is based on Game No.	5	.22	.17	.23	.83	1.33
	6	.06	.12	.72	1.33	.72
	7	.06	.67	1.33	.67	.00
	8	.73	1.45	.73	.00	1.45
	9	.00	.00	.00	.00	.00
	10	.00	.00	.00	.00	2.00

This table shows how far below the value of the game Blue's winnings *can* fall (they may not) if his strategy is based on one matrix while the game is played according to another.

These quantities should have a tendency to increase as one moves to the right, reflecting progressively greater errors in the matrix; and they do have such a tendency, for the average values of the five columns are 0.24, 0.38, 0.39, 0.41, and 0.76.

The lesson that the experiment bears is that reasonable random errors in the matrix—while certainly undesirable—are not necessarily catastrophic. In about half the cases, the greatest possible loss is less than 0.30 in games having values of roughly 3.00 to 4.00; roughly, 10 per cent.

The total situation suggests—at least to the writer—that we have reason to hope for significant results from Game Theory even in fields where the measurement achievements are somewhat substandard.

QUALITATIVE PAYOFFS

A natural continuation of the subject of the last section is that of qualitative payoffs.

A very weak form of measurement is that called *ordering*. Using it, you would simply line up the payoffs—just as you could place in order, according to heat intensity, such items as a piece of ice, a white-hot ingot, a cup of coffee, your blood, and the sun.

If you are unable to make quantitative estimates of the payoffs of interest but you are able to order them, then all may not be lost; for there are game situations which require nothing more than ordering and there are others in which a knowledge of the order will permit you to make partial inferences regarding the optimum strategies or the value of the game.

Suppose you have a game such as the following, in which the payoffs are obscure but which you are able to rate as poor (*p*), fair (*f*), good (*g*), or excellent (*e*). Attack it just as you would if it were expressed in numbers: begin by looking for a saddle-point.

Red

	1	2	3	4	5	Row Min
Blue 1	p	f	g	e	g	p
2	f	f	g	g	e	$f*$
3	p	e	e	e	f	p
4	f	e	f	p	p	p
Col Max	$f*$	e	e	e	e	

Here the greatest row minimum (or maxmin) and the least column maximum (or minmax) are equal. So Blue should play the pure strategy Blue 2, while Red plays Red 1. The value to Blue is *fair*.

We have completely solved this game. Of course, we were lucky, but that isn't reprehensible.

There is a feedback from this type of game to the subject of the last section, namely, accuracy of payoff measurements, which we can state in quite general terms. Suppose that a game has a saddle-point in the box marked by an asterisk:

Red

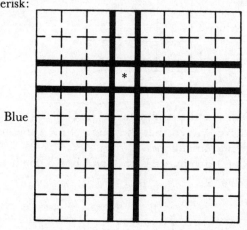

Blue

It doesn't matter where this box is or how large the matrix is.

According to the definition of a saddle-point, the asterisk marks the *largest* number in the *column* in which it appears. It is immediately evident that the other numbers in this column may, without affecting the result, be afflicted by any malady, *provided* only that the asterisked number remains the largest.

Similarly, the asterisk marks the *smallest* number in the *row* in which it appears. The other numbers in this row may be distorted, by accident or design, in any way, as long as the asterisked number remains the smallest.

Finally, the numbers in the matrix which do *not* fall in the same row and column as the asterisk may have any values whatsoever. They are completely unimportant to the play of the game and to its value.

The value of the game depends completely on the asterisk; so the accuracy of measurement of the asterisk determines the accuracy with which the value of the game is known.

In the following game, we can make *some* deductions about optimum play:

		Red 1	Red 2	Red 3	Row Min
Blue	1	p	e	f	p
Blue	2	g	f	g	f^*
Blue	3	p	g	p	p
Col Max		g^*	e	g^*	

The maxmin is *fair* and the minmax is *good;* so there is no saddle-point. But we have learned that the value of the game to Blue lies somewhere between *fair* and *good.*

Moreover, Blue 1 dominates Blue 3, and Red 3 dominates Red 1; so the game reduces to

		Red 1	Red 2
Blue	1	p	e
Blue	2	g	f

The oddment favoring the use of Blue 1 is measured by the difference *good* minus *fair,* and that favoring Blue 2 is measured by the difference *excellent* minus *poor.* So no matter what numerical values should be attached to these words, it is clear that Blue should play Blue 2 more often than Blue 1, and that he should never play Blue 3.

Red's position is a little more fuzzy. He should play Red 1 according to the oddment *excellent* minus *fair,* and Red 2 according to *good* minus *poor;* and it isn't clear which difference is the greater. So all we can say about Red's best mixture is that it is based on Red 1 and Red 2 and that the mix is less extreme than it is for Blue. That is, the odds for Red's mixture are more nearly equal than are Blue's odds. You may convince yourself of the truth of this by assuming some scale of values and trying it out; for instance, if this were the scale,

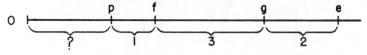

then Blue's odds would be 3:6, and Red's would be 5:4.

EXAMPLE 33. PORTIA

Shakespeare describes in *The Merchant of Venice* a fancy-dress forerunner of the modern shell game.

Portia, a rich heiress, may neither choose who* she would, nor refuse who she dislikes, for husband; for her late father has willed that aspirants to her hand, and to his fortune, must do as follows: Each suitor must face three chests, or caskets, one of which contains a picture of Portia. He may open any one. He who finds the likeness will possess the original. Unlike in the shell game, where the pea is palmed when the stakes become interesting, father actually put a picture in one of the caskets—for he was ever virtuous—but the suitors have no proof of this.

The game has certain side conditions: The suitor must swear, if unsuccessful, to reveal his choice of casket to no one—this protects Portia from a coalition among suitors. He must also swear, if unsuccessful, never to take a wife—the intent of which seems to be to nar-

---

* It has probably been clear for some time that we believe grammar is like money: the principal benefits derive from spending it. However, the present who-whom conflict is strictly between Shakespeare and the grammarians; our role is simply that of the reporter.

row the field to those who regard Portia (and hers) as irreplaceable. On the other hand, helpful hints are provided for those with stomach enough for the game: The chests are of gold, silver, and lead. The one of gold carries the inscription "Who chooseth me shall gain what

many men desire"; that of silver, the inscription "Who chooseth me shall get as much as he deserves"; and that of lead, "Who chooseth me must give and hazard all he hath." At this point, almost every one of Portia's suitors chooseth a powder. The game is evidently a 3 × 3 of the following type:

Father chooses

Gold Silver Lead

		Gold	Silver	Lead
	Gold	P	b	b
Suitor chooses	Silver	b	P	b
	Lead	b	b	P

Here P represents the winning of Portia, to which the suitor must attach a positive value; and b represents permanent bachelorhood.

The suitor may play this as a zero-sum game, or he may attempt to reconstruct father's thinking. Since there is clear evidence that father was an unconventional thinker, the latter approach is not promising.

It is evident, from the symmetry of the matrix, that gold, silver, and lead should be played according to the odds 1:1:1. The value of the game is

$$\frac{P + 2b}{3}$$

It will be a worth-while game for the suitor only if $P + 2b$ is positive. Now P is surely positive and b is probably negative; so the suitor's decision to play the game comes down to this (if he is rational, in the Game Theory sense): He must cherish Portia at least twice as much as he deplores bachelorhood. This information may not make his choice easier, but it sharpens and clarifies the issues.

EXAMPLE 34. THE LADY OR THE TIGER

Frank Stockton's story, "The Lady or the Tiger?" provides the inspiration for this one; we embellish the original, a good 2×2 game, in order to have a 2×3.

The King is disaffected with the Young Commoner because the Princess has accepted him as her lover. To busy idle hands, the King condemns him to open one of two doors. Behind one is a beautiful young lady. Behind the other is a beautiful young tiger. It will be assumed throughout that he prefers the lady to the tiger. (He must marry her.)

You might think this doesn't leave him much of an area for creative thought, but he's awfully interested. It finally comes to him that the King has given him more freedom of action than was intended, for if he opens both doors, he might be able to escape in the resulting confusion.

The Princess, in the meantime, has discovered the King's arrangements—all except that he arranged for her to discover them. She signals to the Young Commoner, indicating the door he should open.

Our boy now has ample room for creative thought. The Princess may share his preference of the lady, if she loves him enough; but if her love is weak, or too intense, she may prefer that he spend the rest of his life with the tiger. The Princess has two strategies: she may direct him to the lady; or to the tiger. He has three: he may follow directions; or he may do the contrary; or he may try to satisfy the tiger with the lady.

Rather than try to guess the precise values the Young Commoner will assign to the alternatives, let's examine the structure of the problem through the graphical method introduced in Chapter 2. We label the payoff boxes alphabetically, for reference:

Princess
tells

		Truth	Lie
	Confidence	*a*	*b*
Commoner's action indicates	Disbelief	*c*	*d*
	Doubt	*e*	*f*

Use two vertical axes for the payoffs (one for Truth and one for Lie). The altitude of an event on these axes is related to the value he assigns it. It is easy to see, qualitatively, how he might rate events (*a*) through (*d*).

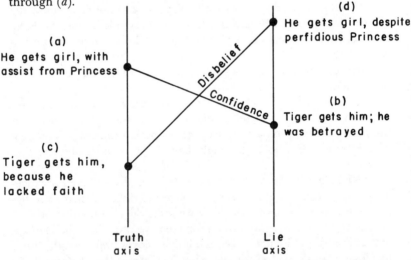

If he were confined to these strategies, Confidence and Disbelief, it is clear that he should use a mixed strategy; the fact that the lines intersect indicates that. [If he rates (*a*) higher than (*d*), the analysis is unaffected, qualitatively.]

Now consider the other two possibilities. He must superpose on the first graph something like this:

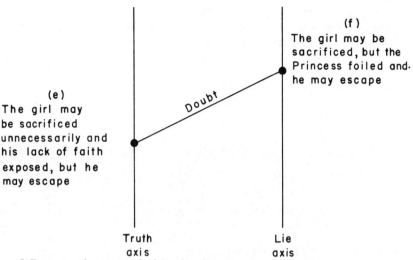

(f)
The girl may be
sacrificed, but the
Princess foiled and.
he may escape

(e)
The girl may
be sacrificed
unnecessarily and
his lack of faith
exposed, but he
may escape

Doubt

Truth
axis

Lie
axis

Whatever the general altitude of these events, the line surely slopes upward (going from Truth to Lie).

The knotty part of the problem comes when the two graphs are put together. The critical issue will be: Does the Doubt line pass above or below the intersection of Confidence and Disbelief. If it passes above, he should mix Confidence and Doubt; if below, he should mix Confidence and Disbelief. This does not depend on how much above or below it passes, nor on its slope—provided it doesn't get so far above the other lines as to be completely isolated, in which case Doubt would become a saddle-point solution. The representative situations look like this:

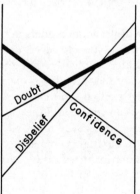

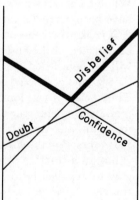

In order to estimate the oddments for each strategy, it would be necessary for us to put ourselves in the Young Commoner's place— but let's leave the suffering to him.

GAMES PLAYED ONLY ONCE

The fact that we have avoided all discussion of one subtle point may cause unhappiness among lay and professional readers alike— but for diametrically opposite reasons! It seems that the least we should do is to bring the subject out into public view, even though it may not lead to dancing in the streets, in the end.

Lay readers may believe it self-evident that Game Theory is inappropriate for games that can be played only once, for the very concept of a mixed strategy seems to require that there be some repetition of play. These readers can point to certain of our examples, such as The Lady or The Tiger, as obviously being nonrecurring affairs. While being very pleasant about it, it will be clear that they feel we have been carried away by it all.

It is likewise conceivable that some of our professional readers may be alarmed at the absence of discussion on this point. However, they will be concerned because we have not stated explicitly that the theory should, strictly, only be applied to games that are played once. They perhaps will look askance at examples such as The Daiquiris (and, worse, The Huckster) which obviously can recur; and they too may hint, pleasantly, that we may have been carried away by it all.

It takes something less than second sight to see that such a subject is not an attractive place for a long excursion, for it would be very easy to step well beyond the limits appropriate to a primer. We shall therefore confine the tour to points from which the lay reader may glimpse—we trust—solid ground through the fog.

We hope that the lay reader will be able to accept the flat statement that the strictly mathematical aspects of Game Theory were developed with reference to a single play of a game. Once that is accepted, there is no great difficulty in believing it to be applicable to games that can be played but once. As we have developed the Theory of Games for you here, the implication has been strong that we were talking about repeatable events. The point we are now making is that it was convenient for us to talk that way (and scientifically justifiable, too, we believe), but it was not necessary to do it that way; for the frequency aspects were not required in the delicate mathematical inquiries which created the Theory.

The above requires a little faith, at least in the reliability of the reporter. The following kind of argument, on the other hand, may make a direct appeal to you: Consider a nonrepeatable game which is terribly important to you, and in which your opponent has excellent human intelligence of all kinds. Also assume that it will be murderous if this opponent *knows* which strategy you will adopt. Your only hope is to select a strategy by a chance device which the enemy's intelligence cannot master—he may be lucky of course and anticipate your choice anyway, but you have to accept some risk. Game Theory simply tells you the characteristics your chance device should have.

You may also adopt the viewpoint that you will play many one-shot games between the cradle and the grave, not all of them being lethal games, and that the use of mixed strategies will improve your batting average over this set of games.

The professional Game Theorist has two viewpoints here that may be interesting to you. One is that, since the theory was developed explicitly in terms of one play of a game, he isn't certain that this is the theory he would have developed if he had been told the game was going to be played exactly eleven times, say. It seems to the writer that this attitude reflects good science, but that it is unnecessarily pure for the practitioner who doesn't have Eleven-game Theory at hand, and who can't wait.

In games against Nature, however, the attitude of the purist cannot be shrugged off so easily, for here a long sequence of repetitions of a game may contain information regarding the pattern of Nature's grand strategy which should not be ignored. In fact, the domain of the Game Theorist here blends into that of the Statistician, who himself has a bag full of scientific tricks which should be brought to bear on the problem.

SYMMETRIC GAMES

Another type having special properties is the *symmetric* game. A game is symmetric when the rows and columns of its matrix are related in this manner: The sequence of payoffs in each row is the same as the sequence in the corresponding column, except that the signs are reversed. As one consequence of this definition, the elements on the main diagonal must be zeros. Thus

0	2	3
−2	0	−6
−3	6	0

is a symmetric game. Note, for example, that the third row, −3, 6, 0, and the third column, 3, −6, 0, have the required property, as do the other pairs of rows and columns.

There are two useful facts about a symmetric game. One is that it is a fair game; and since we know its value is zero, we do not have to compute it. The other is that both players may use the same grand strategy.

Every game may be transformed into another larger game which is symmetric; this may be desirable for computational reasons. In order to do this, it is necessary that the original game have a positive value, but we can always achieve this by adding a large constant to each payoff, which will not affect the strategies—and it can be subtracted from the value, at the end.

To exhibit the method, consider a game and two auxiliary matrixes of 1's:

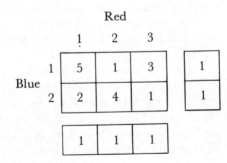

Red

	1	2	3
Blue 1 | 5 | 1 | 3 |
Blue 2 | 2 | 4 | 1 |

call these G, V, and H, for Game, Vertical, and Horizontal matrixes. Then build a supermatrix of this form:

0	G	$-V$
$-G^*$	0	H^*
V^*	$-H$	0

Here, an asterisk is a command to change rows into columns; a minus sign means to change the signs of all elements; and 0 stands for all-zero matrixes. Substituting for these symbols their values, you get, finally,

Red*

	1	2	3	4	5	6
1	0	0	5	1	3	−1
2	0	0	2	4	1	−1
3	−5	−2	0	0	0	1
4	−1	−4	0	0	0	1
5	−3	−1	0	0	0	1
6	1	1	−1	−1	−1	0

Blue* (at left, rows 3–4)

which is a symmetric game. Both 'players' of the 6 × 6 may use the same grand strategy. (The asterisks on Blue* and Red* just serve to distinguish these fictitious players from the real ones of the original game.)

Now, if you can find a grand strategy for this game in which the last column is one of the active pure strategies, then the strategies of this game will be related to those of the original game according to this scheme:

	Blue		Red			Value of
Strategies of Original Game	1	2	1	2	3	Game
Strategies of Symmetric Game	1	2	3	4	5	6

That is, the oddments for the various pure strategies of Blue and Red will be given by the oddments associated with the first two groups of pure strategies in the symmetric game, and the value of the original game will be proportional to the oddment for the last pure strategy. In the present example a solution to the symmetric game is 3:2:0:2:3:11, which yields this for the original game:

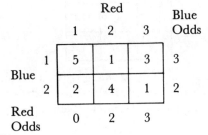

		Red 1	Red 2	Red 3	Blue Odds
Blue	1	5	1	3	3
	2	2	4	1	2
Red Odds		0	2	3	

Its value is $1\frac{1}{5}$.

LINEAR PROGRAMMING

We have observed, in the examples which have appeared throughout the book, that Game Theory may be relevant to a wide variety of fields of human activity. It should not be completely unanticipated then to discover that it is also related to theoretical developments in various fields. The relationship is sometimes at the conceptual level; at other times, it is purely formal, as when the mathematics is similar. In either case the fact is beneficial to all concerned, for the interplay of ideas and methods thereby enriches all fields.

Simply as an example of this interplay, we shall exhibit one case: it is that aspect of scientific management known as Linear Programming. The United States Air Force has pioneered in this field, attemping to develop a rational planning and procurement scheme which will yield, at various future dates, the mixture of goods and sources needed for its tasks; and to achieve this most economically. Computationally, problems of this type may be brought to the form of a game, as the following example illustrates.

EXAMPLE 35. THE DIET

John, following a familiar pattern, has discovered a new diet; this one is based, primarily, on meat. He must eat 1 ounce of fat with each 3 ounces of lean. Since cost is an important aspect of an all-meat diet, he decides to find out how cheaply he can buy a nominal quantity—at least 10 pounds, say—of the right mixture of fat and lean, in order to try it out.

Dantzig's Butcher Shop carries three kinds of meat which John is willing to eat. These contain 1, 5, and 10 ounces of fat per pound and retail at $1.35, $0.85, and $0.65, respectively. These facts may be summarized in three matrixes as follows (the units are ounces and cents, respectively):

		Fat	Lean	Price
	1	1	15	135
Meats	2	5	11	85
	3	10	6	65

Quantities Desired	40	120

We shall identify these as the M, P, and Q matrixes (for meats, prices, and quantities).

We now form a new matrix—a whopper—in which the foregoing matrixes are the elements, thus:

0	M	$-P$
$-M^*$	0	Q^*
P^*	$-Q$	0

Here, a minus sign means that each element in the submatrix must be prefaced by a minus sign; an asterisk means that the rows of the submatrix must be changed to columns; and 0 stands for a sub-matrix made up of zeros. Substituting for the symbols M, P, and Q, the new matrix expands to

Red

	1	2	3	4	5	6
1	0	0	0	1	15	−135
2	0	0	0	5	11	−85
3	0	0	0	10	6	−65
4	−1	−5	−10	0	0	40
5	−15	−11	−6	0	0	120
6	135	85	65	−40	−120	0

Blue (label at left, rows 1–6)

As an old hand at this sort of thing, you will instantly recognize this as a game—simply because every matrix may be so regarded. This one happens to be a 6 × 6; and nicely symmetric, too, which ensures that it is a fair game and that a grand strategy which is suitable for one player is also suitable for the other. The thing that one could not anticipate was that its solution is intimately connected with the dietary problem.

Now every game has a solution, but some linear programming problems do not. If the linear programming problem does have one, then (it turns out) there must be a grand strategy for the game in which the last row (or column; 6 here) plays an active role: i.e., its oddment is not zero. When this is true, the linear programming solution is found by dividing the oddments for each of the first group of pure strategies—1, 2, and 3 here—by the oddment for the last strategy —6 here.

In the present example, we find that 0:120:0:0:85:11 are the odds according to which the game should be played. The last oddment being 11—which is not zero—the linear programming problem has a solution.

In fact, John should buy 9/11, 120/11, and 9/11 pounds, respectively, of the three meats; i.e., he should buy 10 10/11 pounds of the $0.85 variety and shun the others. Thus for about $9.27 he gets 120 ounces of lean and over 54 ounces of fat. This excess of fat, over his 40-ounce requirement, is necessary waste; it would cost more to buy precisely the correct mixture. (Prizes will *not* be offered for easier ways to solve this particular dietary problem.)

NON-ZERO-SUM GAMES

We shall do no more than indicate the nature of these games. To do more would take us into deep water mathematically; moreover, into water riled by controversy.

In zero-sum games the payoffs represent strictly an exchange of assets; one player wins the quantity that the other loses. We have compromised this principle somewhat in games played against Nature (used as examples here and there), where we have computed strategies for the personal player as if he were playing against an opponent who shared his valuation of the payoff matrix and who was capable of making intelligent countermoves. This led the personal player to overconservative strategies; he failed to win as much as was available when playing against ironclad stupidity. In principle, there is a satisfying way to improve this: simply discover Nature's grand strategy (this may be possible in specific instances), and then capitalize on the weakness of Her method.

The non-zero-sum game has in it an element similar to Nature. Here the winnings of one player are not necessarily the losses of the other, for they may both win or both lose; so, to preserve some part of the form of our analysis, we may think of Nature as being in the game too, absorbing the valuables lost by the players and supplying their winnings. Such a game is described by two matrixes, and may have this appearance:

	Nature (R)					Red		
	1	2	3			1	2	3
Blue 1	1	6	2		1	6	0	3
Blue 2	3	0	3	Nature (B) 2		5	4	3
Blue 3	4	2	0		3	0	1	4

Thus Nature has a dual personality; one aspect is relevant to Blue and one to Red.

If Blue considers only the left-hand matrix and regards it as defining a conventional two-person, zero-sum game, he will adopt the grand strategy 4:6:1 and his winnings will average 2⁴⁄₁₁, the payee being Nature (R). The right-hand matrix could similarly lead Red to use 0:1:3 and to losses of 3¼, collected by Nature (B). Nature (B) and Nature (R) pool their winnings and losses, if you desire it.

Now, in a one-person game against Nature, Blue could win more than 2⁴⁄₁₁ by studying Her behavior and capitalizing on his knowledge; but in the present game, Nature (R) is a wolf in sheep's clothing, for Red lurks in the background and controls Her strategy. There is, in effect, a coalition in which Red is the general and Nature (R) is the banker; Blue and Nature (B) are similarly related. So if Blue departs from the grand strategy 4:6:1 in order to win a little more, he may get clobbered by an attentive Red.

But there is still more to it: Suppose Blue elects to use the strategy 1:0:0 and that Red chooses 0:1:0. The payoff to Blue now becomes 6, instead of 2⁴⁄₁₁; and Red's losses become 0, instead of 3¼. So both players are vastly better off, in that Blue wins as much as possible and Red loses as little as possible. *It is evident that Blue and Red can, with mutual profit, form a coalition against Nature, in order to win from Her as much as possible.*

This leads us squarely into difficulties. There is the problem, classic among thieves, of how to divide the loot. If the Coalition can earn more than the total the individuals can earn by playing independently, how should the excess be divided? Or, what system of side-payments should be made among the players to ensure that the Coalition will prosper—that its income will be high and the members loyal? When will a player accept a less favorable payoff in order to penalize the other player? Much work is being done in this field in an effort to invent satisfactory criteria which will guide one to sound strategies of over-all behavior—strategies of play and systems of side-payments. The abstract work is in serious need of guidance, however, from practical data, data drawn from real-life situations, and from laboratory experiments. The subject, though fascinating, is clearly beyond the modest limits of this book.

CONCLUSION

We are now in a position to do what we were not able to do in the beginning: namely, to give a fairly concise statement of what Game Theory is, just as we did in Chapter 1 for the Theory of Gravitation.

For we now have a large technical vocabulary, and sufficient familiarity with the mechanism of analysis and the formulation of problems, to understand a series of statements which would in the beginning have been intolerably technical.

The following description of the theory is aimed only at the material which is spanned by this book. Some of the items are subject to generalization, as to infinite games, non-zero-sum games, and n-person games.

The Theory of Games is a method of analyzing a conflict, according to the following abstraction: The conflict is a situation in which there are two sets of opposing interests; it may be regarded as a game between two players, each of whom represents one set of interests. Each player has a finite set of strategies from which he may, on any given play of the game, choose one. The total assets of the players are the same at the end of any one play of the game as at the beginning. Each player wishes to pursue a conservative plan which will maximize his average gains; these maximum average gains, called the value of the game, may be calculated. Each player can, through proper play, be sure that he will receive the value of the game; to ensure this, he must choose his strategy properly—and a method exists for deciding which strategy to choose. There are various corollaries: he *may* have to conceal from the enemy his choice of strategy on each particular play; it is never necessary for him to conceal more than his current choice; etc.

Mathematics has been called many things. Among others, the queen of the sciences and the handmaiden of the sciences. Therefore, while there seems to be some doubt regarding its precise social position, there is general agreement regarding its gender. In fact, all sciences probably look pretty feminine to the men who attend them: They obviously find the work desperately attractive, and the history of scientific investigation is full of evidence that she whom they court is changeable and capricious; in short, something of a hussy.

Moreover, the way she responds to a push, or to guidance, is usually astonishing. She moves freely, but frequently in an unexpected direction. Time and again men have spent their lives trying to move her in a specified direction, quite without success. Viewed narrowly, their work would be classed as a failure. But, since science usually moves somewhere when pushed, these failures often are the bases of completely unexpected successes in other fields.

Many examples of this—some of which are famous—could be adduced. We will recount one, briefly, just to show how it goes: Sunspots—the great fire tornadoes on the surface of the sun—occur periodically. The streams of particles and energy released have some effect on the earth; long-distance radio reception, for instance, is affected. It is conjectured that sunspots affect the earth's weather, too, but the form of the dependence is obscure.

It occurred to an astronomer—A. E. Douglass—that the annual growth of trees, as reflected by the rings of trees, must constitute a

long series of rainfall records; and that if the weather is affected by sunspots, there should be a record of sunspot activity impressed on the tree-ring sequence.

He spent about forty very intensive years working on the problem. The main effect he was seeking was clearly evident in the records, but the detailed effects were complex and controversial to a high degree; so much so that the entire analysis was at times under a cloud. However, he constructed—and this was of only incidental importance to the initial problem—a very firm and precise chronology for trees, extending back in time for two thousand years or more. It is now possible, through use of this tree-ring calendar, for archaeologists to date the ruins of early civilizations very accurately—in fact, to the year—from a fragment of wood found in a building or a piece of charcoal from a campfire.

Game Theory was originally developed—by a mathematician—with a view toward certain problems of economic theory. The initial reaction of the economists to this work was one of great reserve, but the military scientists were quick to sense its possibilities in their field, and they have pushed its development. It seems, however, to have been taken over by the economists once more, along with other quantitative methods which are rapidly changing the nature of economic theory and practice. One of the most important of these methods is linear programming, which has rapidly grown to be such an important tool in industrial planning that some firms no longer can conceive how they might operate without it, and devote the majority of their large computer budgets to its use. The general notions of Game Theory have been at the forefront of a revolution in the theory of statistics, where a great unification of apparently disparate items of knowledge is taking place. The concepts are now being explored in relation to social science, where they may shed light on certain interactions that occur among people. It is difficult to predict where it may turn up next.

While there are specific applications today, despite the current limitations of the theory, perhaps its greatest contribution so far has been an intangible one: the general orientation given to people who are faced with overcomplex problems. Even though these problems are not strictly solvable—certainly at the moment and probably for the indefinite future—it helps to have a framework in which to work on them. The concept of a strategy, the distinctions among players, the role of chance events, the notion of matrix representations of the payoffs, the concepts of pure and mixed strategies, and so on give valuable orientation to persons who must think about complicated conflict situations.

CHAPTER 6

General Method of Solving Games

The game-solving methods described in this book have—except for the approximate method introduced on page 182—a common weakness: their applicability depends on the nature of the solutions. For instance, we have a method to find a saddle-point solution *if* there is a saddle-point, another method to find the all-strategies-active solution *if* all pure strategies are in fact active, and so on. Thus the methods are conditional. By running through a sequence of methods and sub-games—often a long sequence—we eventually find a solution; but until we find it, we just eliminate bad guesses. Clearly, a general constructive method of solution would be preferable.

In this chapter we shall describe a method, called the *pivot method,* which is powerful enough to ferret out all solutions, and which is efficient enough to be practical; that is, it usually reaches the exact solution in a few steps. The method is more complicated—particularly to describe—than the methods discussed earlier, but we believe the careful reader can follow the presentation.

The problem of finding the solutions to a game may be likened to the problem of a squirrel that seeks the highest tip of a tree. If he ascends by small steps and explores every branch, he will certainly find the highest tip; and if the tree has a flat top, he will find all the topmost tips—the latter situation corresponds to a game with more than one basic solution. This plan has the virtue of simplicity, but it clearly involves a lot of detailed work. On the other hand, the squirrel may have a plan to get to the top by leaps, which is more complicated but will eliminate much of the detailed work. If he wants to find the topmost tips, he may have a further plan, once he reaches the top, to leap from tip to tip. This would be an additional complication in the over-all plan, but it would be less work than to climb the tree repeatedly— moreover, he might not know where to start the new climbs because he might have missed important branches during earlier ascents.

The arithmetic labor of solving games is so great that we imitate the sophisticated squirrel: we seek a solution by leaps and, if we are interested, leap from solution to solution. This puts strain on our heads, but it relieves our backs.

FIRST EXAMPLE

Consider the game encountered on page 95 (henceforth we shall not put the entries of a game matrix in boxes):

		Red		
		1	2	3
	1	6	0	3
Blue	2	8	−2	3
	3	4	6	5

We found its value to be 4⅔, the optimal strategies to be Blue 0 : 1 : 5 and Red 2 : 1 : 0.

Step 1. The first step in the new process is to search the original game matrix for negative numbers. If any are present, a positive number must be added to every number in the matrix. Any positive number will serve provided it is large enough to eliminate all the negative numbers. In the present case the number + 2 will suffice, and the game becomes

		Red		
		1	2	3
	1	8	2	5
Blue	2	10	0	5
	3	6	8	7

You will recall that adding a constant to a game matrix does not affect the choice of strategies, but it does affect the value of the game; so we shall find that the value of the game we are now considering is 4⅔ + 2 = 6⅔. Therefore we will have to remember, at the end of the work, to subtract 2 from the value of the present game in order to arrive at the value of the original game.

Step 2. The second step is to augment the present game matrix with a border of 1's, −1's, and 0, and to provide an auxiliary number, D (equal to 1 initially), all in the following pattern:

FIRST SCHEMA

Red

		1	2	3		
	1	8	2	5	1	
Blue	2	10	0	5	1	
	3	6	8	7*	1	
		−1	−1	−1	0	$D = 1$

This is in general called a schema; in particular, the First Schema. The asterisk will be explained presently.

We shall give a prescription for a sequence of schemata such that, in the final one, Blue's optimal strategy will be determined by numbers that appear in the bottom border, Red's by numbers in the right border, and the value of the game by the border number in the corner and by D. We will know that a schema is the final one, and hence that it has these felicitous features if no *border* element is negative—provided we have made no errors.

Step 3. The next step is to select an appropriate cell, which will be known as a *pivot,* from those of the *game matrix* section of the schema. This pivot, by its position and value, will affect all the values of the next schema. The game-solving process consists mostly of finding and using suitable pivots.

To find the pivot we must consider several trios of numbers: the potential pivot itself—call it p—and the border elements in the row and column of the potential pivot—call them r and c.

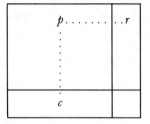

The rules are

Step 3.1. The potential pivot p must be positive.

Step 3.2. The number at the foot of the column c must be negative.

Step 3.3. Compute for each potential pivot the quantity

$$- \frac{r \times c}{p}$$

called the *pivot criterion*. This quantity is never negative, because p is positive, c is negative, and r is never negative (a fact, not a rule).

Step 3.4. Underline the *smallest* value of the criterion in each column.

Step 3.5. Mark with an asterisk the *largest* of the underlined values. The value of p that corresponds to the asterisked value is the pivot P.

In the First Schema, all c's are negative; so all nonzero p's must be considered. However, the c's and r's are all -1's and 1's, respectively; so the criterion becomes

$$- \frac{1 \times (-1)}{p} = \frac{1}{p}$$

and the smallest values can be found by inspection. In our example, we inspect this display:

⅛	½	⅕	
$\underline{1/10}$		⅓	
⅙	$\underline{⅛}$	$\underline{½}$*	

The smallest values in the columns are ⅒, ⅛, and ½, and ½ is the largest. Therefore the pivot is $P = 7$, in the third row and third column. It is marked by an asterisk in the First Schema.

Step 4. The numbers in the next schema are found as follows:

Step 4.1. The number which corresponds to the pivot is the value D of the preceding schema.

Step 4.2. The numbers which correspond to those of the preceding pivot row are the same as before—except for the pivot value itself.

Step 4.3. The numbers which correspond to those of the preceding pivot column are the same as before but of opposite signs—except for the pivot value itself.

Step 4.4. A number which corresponds to one—call it N—which is neither in the pivot row nor in the pivot column is computed from

$$\frac{N \times P - R \times C}{D}$$

where R and C are the numbers which have rows and columns in common with P and N, as indicated schematically:

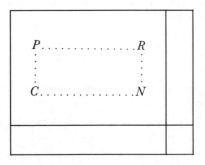

Step 4.5. The next value of D is the pivot value P of the preceding schema.

Let us apply these five steps progressively to determine the next schema for the example.

By *Step 4.1:*

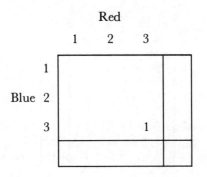

By *Step 4.2:*

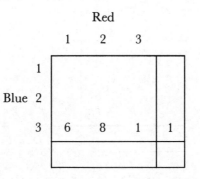

By *Step 4.3:*

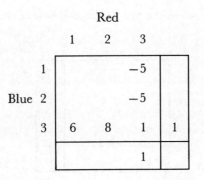

By *Steps 4.4* and *4.5:*

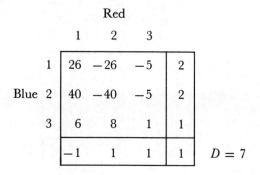

In case some reader is baffled by the symbols that appear in *Step 4.4*, we show the details for two cases. The number -26 in the first row and second column is determined from these:

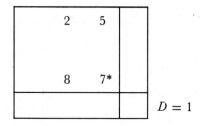

by

$$\frac{N \times P - R \times C}{D} = \frac{2 \times 7 - 8 \times 5}{1} = -26$$

and the number -1 in the fourth row and first column is determined from these:

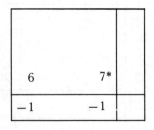

by
$$\frac{(-1) \times 7 - 6 \times (-1)}{1} = -1$$

Step 5. The numbers along the left and upper edges of the First Schema are the names of the Blue and Red strategies. The final part of each pivot operation is

Step 5.1. Exchange the Blue strategy name that is at the left of the pivot row, if one is there, with the name that is below the pivot column, if one is there.

Step 5.2. Similarly, exchange the Red strategy name that is above the pivot column, if one is there, with the name at the right of the pivot row, if one is there.

The 'if one is there' caveats are probably mystifying if this is your first encounter with a schema because in the First Schema there are surely names to the left of the pivot row and above the pivot column and none elsewhere, but this will not be so for later schemata. For instance, we could have a schema, with pivot $P = 25$, with these features (among others):

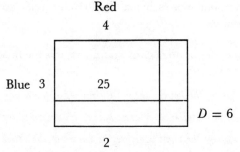

Here we need the caveats in order to produce a next schema having these features (among others):

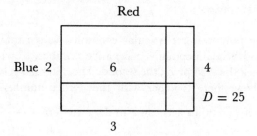

In the present example, the complete next schema, the second, is

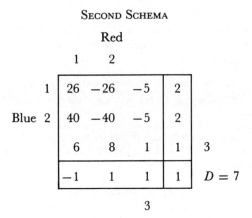

SECOND SCHEMA

Red

Step 6. Check the border of the new schemata for negative numbers.

Step 6.1. If it contains one or more negative numbers, we must compute at least one more schema. To do so, go again to *Step 3* and do *Steps 3, 4, 5,* and *6.*

Step 6.2. If it contains no negative numbers, it is the final schema. It is interpreted as follows:

The numbers along the lower edge are the names of the active Blue strategies.

The border numbers adjacent to Blue names are the corresponding Blue oddments.

The numbers along the right edge are the names of the active Red strategies.

The border numbers adjacent to Red names are the corresponding Red oddments.

The value of the game is found by dividing the auxiliary number D by the border number in the corner.

Returning now to our particular example, we see that the border of the Second Schema contains a negative number; so we must go to *Step 3* and seek the next pivot, the second. The potential pivots are three numbers, shown here together with the border numbers needed to evaluate their potentials according to the criterion $-\dfrac{r \times c}{p}$:

26	2
40	2
6	1
−1	

The values of the criterion are

.077	
.050*	
.167	

We see that the value corresponding to the 40-cell is the smallest; so
$P = 40$. It is marked by an asterisk in the Second Schema.

Applying the rules of *Steps 4* and *5*, we build up the Third Schema
as follows:

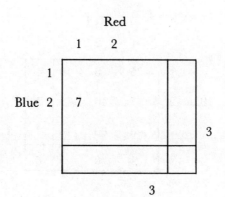

Red

	1	2	
1			
Blue 2	7		3

3

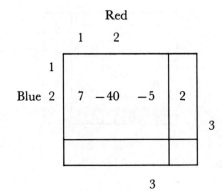

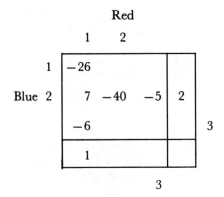

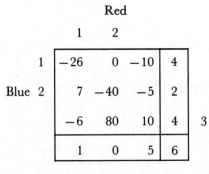

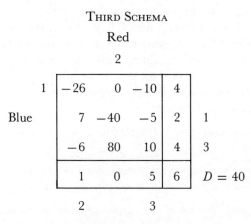

THIRD SCHEMA

Red

2

		2		
1	−26	0	−10	4
Blue	7	−40	−5	2	1
	−6	80	10	4	3
	1	0	5	6	$D = 40$

Since the last row contains no negative numbers, we may interpret this schema according to the instructions in *Step 6.2*. An optimal strategy for Blue is a mixed strategy based on Blue 2 and Blue 3 in the ratio 1:5. Red should use Red 1 and Red 3 in the ratio 2:4 or 1:2. We represent these by Blue 0:1:5, Red 1:0:2. The value of this game is $^4\!\!\%= 6\frac{2}{3}$— as we anticipated, 2 greater than the value $4\frac{2}{3}$ of the original game.

The rules for the solution of games given above are the most efficient ones known to the writer, but there is no guarantee that they will lead to the least number of pivots. They happened to be efficient in the example just given, where two pivots led to the two-strategy solution, but one may not always be so lucky.

BASIC SOLUTIONS

We discussed basic solutions earlier (page 191), but we did not have a powerful method of finding them. We are about to have one.

If a game has more than one basic solution, that fact will manifest itself through the occurrence of one or more zeros in the border of the final schema. We can find another basic solution by proper choice of pivot in a row or column which has a zero border element.

We can find all basic solutions by pivoting through all sequences of rows and/or columns which have zero border elements, but we do not use the same pivot twice in succession.

Thus if two columns—call them A and B—have zero border elements, we pivot in the sequence A, B, A, B, ... We shall find a basic solution after each pivot operation. We also try the sequence B, A, B, A, ... If there are two equally preferable pivots in a row or column, we try sequences based first on one, then on the other. To examine exhaustively the possible sequences apparently can be a major problem, if there are many border zeros.

A sequence terminates if we find an old basic solution, or if there is no suitable pivot.

An old basic solution may be recognized without computation by the recurrence of a set of Blue and Red strategy names adjacent to the border.

The order and position of the names adjacent to the border are not significant; it is an old solution if the names are the same.

Step 7. The pivot selection rules are

Step 7.1. If the border element of a column is 0, choose as pivot P that positive element p for which ratio r/p is smallest.

Step 7.2. If the border element of a row is 0, choose as pivot P that negative element p for which the ratio c/p is largest (i.e., closest to zero).

Step 8. Go to *Step 5.* After transferring the strategy names as required by this pivot operation, compare the border sets of Blue and Red names with the set found for the first basic solution (or with the sets found for all previously found basic solutions).

Step 8.1. If the present border set of Blue and Red names is novel, go to *Step 4* and calculate the new schema, but modify the rules as follows if the pivot P is negative: Make the old value of D negative, and change the signs of the elements discussed in *Steps 4.2* and *4.3.*

Step 8.2. If the present border set of Blue and Red names is not novel, terminate this sequence of basic solutions.

Let us try all this on an example. On page 229 we found a basic solution, represented by the following schema, for the game formulated on page 219:

SCHEMA: SOLUTION 1

Red

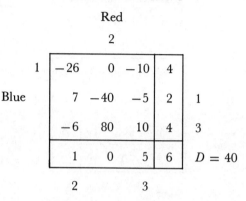

This has a zero border element in the second column, so there is at least one additional basic solution. There is one positive number p in the column. According to *Step 7.1*, that must be the pivot; i.e., $P = 80$. We transfer strategy names as directed by *Step 8* and find

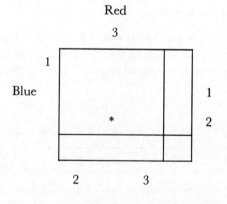

This set of border names is new (a Red name has changed); so we are about to find a new solution. *Step 8* instructs us to calculate the rest of the new schema as in *Step 4*. It turns out to be

Schema: Solution 2

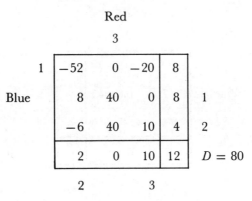

Red

3

1	−52	0	−20	8
Blue	8	40	0	8	1
	−6	40	10	4	2
	2	0	10	12	$D = 80$

2   3

Solution 2 requires Blue to mix Blue 2 and Blue 3 according to the oddments 2 and 10, or 1 and 5, whereas Red is to mix Red 1 and 2 as 8 is to 4, or as 2 is to 1. We write Solution 2 as Blue 0:1:5, Red 2:1:0. Solution 1 was Blue 0:1:5, Red 1:0:2.

We conclude that Blue has one basic strategy and that Red has two. It follows that Red has infinitely many optimal strategies, because he may use any mixture of his two basic strategies.

SECOND EXAMPLE

We include one more worked example to provide the reader with a lot of arithmetic with which to test his comprehension of the solution process. We shall use the game on page 193, Exercise 6, and we shall seek all basic solutions.

Red

		1	2	3	4
	1	16	−8	9	−3
	2	−20	4	9	−3
Blue	3	25	1	18	−6
	4	−11	13	−18	6

We add 20 to each element so as to eliminate the negative numbers, and we provide the border of 1's, -1's, and 0, as well as the auxiliary number $D = 1$, needed to produce the first schema:

SCHEMA 1

Red

	1	2	3	4	
1	36	12	29	17	1
2	0	24	29	17	1
3	45	21	38	14	1
4	9	33	2	26*	1
	-1	-1	-1	-1	0

(Blue labels rows 1–4; $D = 1$)

All numbers in the game matrix sector of this schema except the 0 must be considered as pivots because they are positive and the lower border numbers are negative; so one must calculate the value of the criterion

$$- \frac{r \times c}{p}$$

for each. This becomes just

$$\frac{1}{p}$$

because all r's and c's are 1 and -1, respectively.

$\frac{1}{36}$	$\frac{1}{12}$	$\frac{1}{29}$	$\frac{1}{17}$
	$\frac{1}{24}$	$\frac{1}{29}$	$\frac{1}{17}$
$\frac{1}{45}$	$\frac{1}{21}$	$\frac{1}{38}$	$\frac{1}{14}$
$\frac{1}{9}$	$\frac{1}{33}$	$\frac{1}{2}$	$\frac{1}{26}$*

The smallest value of the criterion in each column is underlined and the greatest of these, $\frac{1}{26}$, is marked by an asterisk; so the pivot is $P = 26$.

We proceed to Schema 2 by *Step 4* as follows:

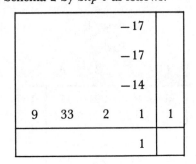

Here the pivot is equal to D, the column numbers have changed signs, and the row numbers have been copied. The remaining numbers of the new schema are computed by the formula

$$\frac{N \times P - R \times C}{D}$$

For instance, that number in the first row and second column is

$$\frac{12 \times 26 - 33 \times 17}{1} = -249$$

The complete schema is

SCHEMA 2

Red

		1	2	3		
Blue	1	783	−249	720*	−17	9
	2	−153	63	720	−17	9
	3	1044	84	960	−14	12
		9	33	2	1	1
		−17	7	−24	1	1

4     $D = 26$

4

The potential pivots in Schema 2 are the positive numbers in columns that have negative border numbers, namely,

783	720	9
	720	9
1044	960	12
9	2	1
−17	−24	

Here we have reproduced the border elements needed to evaluate the potential pivots according to the criterion

$$- \frac{r \times c}{p}$$

For instance, the criterion for judging the number 1044 is

$$- \frac{12 \times (-17)}{1044} = .20$$

(Approximate calculations suffice, so long as choices can be made accurately.)

.20	.30*	
	.30	
.20	.30	
1.9	12	

The smallest values are approximately .2 in the first column and .3 in the third. The larger is .3; so the pivot corresponds to it. In fact we have three equally preferable pivots: 720, 720, and 960. We select arbitrarily the 720 in the first row and third column and mark it by an asterisk in Schema 2. The other two pivots are not of interest because we are going to climb the game tree just once.

We proceed to the next schema, which is

SCHEMA 3

Red

		1	2				
		783	−249	26	−17	9	3
	2	−25,920	8,640*	−720	0	0	
Blue	3	0	11,520	−960	240	0	
		189	933	−2	29	27	4
		252	−36	24	12	36	$D = 720$

1   4

Here the potential pivot must be a positive number in the second column, the only one with a negative border number. The elements relevant to the choice are

8,640	0
11,520	0
933	27
−36	

The criterion $-r \times c/p$ is 0 in two instances because two of the c's are 0; so 8,640 and 11,520 qualify as pivots. We choose the former and calculate the next schema.

Schema 4

(Solution 1. blue 7:1:0:4, red 0:0:1:3)

Red

	1					
	432	249	63	−204	108	3
	−25,920	720	−720	0	0	2
Blue						
3	414,720	−11,520	0	2,880	0	
	35,856	−933	909	348	324	4
	1,728	36	252	144	432	$D = 8,640$

| | | 2 | 1 | 4 | | |

No border element is negative; so we have reached a solution.

We see that Blue should play Blue 1, 2, and 4 according to the odd-ments 252, 36, and 144, or, removing the common factor 36, according to the oddments 7, 1, and 4. Similarly, Red should play Red 2, 3, and 4 according to the oddments 0, 108, and 324, or 0, 1, and 3. The value of the game is $8,640/432 = 20$, which is equal to the constant added at the beginning of the work; so the original game is a fair game.

Is Blue 7:1:0:4, Red 0:0:1:3 a solution? Or have we made errors? It is easy to find out. The result of using Blue 7:1:0:4 against the strategy Red 1 is (from the original matrix)

$$\frac{16 \times 7 + (-20) \times 1 + 25 \times 0 + (-11) \times 4}{7 + 1 + 0 + 4} = \frac{48}{12} = 4$$

Against Red 2 it is

$$\frac{(-8) \times 7 + 4 \times 1 + 1 \times 0 + 13 \times 4}{12} = \frac{0}{12} = 0$$

Against Red 3 it is

$$\frac{9 \times 7 + 9 \times 1 + 18 \times 0 + (-18) \times 4}{12} = \frac{0}{12} = 0$$

Against Red 4 it is

$$\frac{(-3) \times 7 + (-3) \times 1 + (-6) \times 0 + 6 \times 4}{12} = \frac{0}{12} = 0$$

Thus Blue wins the value of the game, or more, against any defense.

The Blue solution is valid, apparently. Can we assert that it is? The shocking fact is we cannot: It may be that we have miscalculated the value of the game. If the true value is ½, say, rather than 0, the preceding test has mislead us. To be sure that the Blue strategy and the estimate of the value of the game are both correct, we must complete the test by evaluating Red's presumed-optimal strategy, namely, Red 0:0:1:3. Red's average result against Blue 1 is

$$\frac{16 \times 0 + (-8) \times 0 + 9 \times 1 + (-3) \times 3}{0 + 0 + 1 + 3} = 0$$

and it is the same against Blue 2, 3, and 4. Since both players average the same when using the mixtures we have proposed, and do at least as well against any pure strategy, we can now assert that the value is 0 and that we have found a solution.

Does the game have more than one basic solution? Apparently it does because there are zero border elements. These occur in rows—the second and third—so, according to *Step 7.2*, we must look for negative pivots and make *D* negative. With this modification Schema 4 becomes

Schema 4'

Red

	1					
	432	249	63	−204	108	3
	−25,920*	720	−720	0	0	2
Blue 3	414,720	−11,520	0	2,880	0	
	35,856	−933	909	348	324	4
	1,728	36	252	144	432	$D = -8,640$
	2	1	4			

We will seek pivots in rows 2 and 3 in the sequence 2, 3, 2, 3, . . . , until it terminates. Considering row 2, the potential pivots are $-25,920$ and -720. The criterion c/p has the values $1,728/(-25,920) = -.07$ and $252/(-720) = -.35$. The first value is closer to zero than is the second; so $P = -25,920$. Transferring strategy names about this pivot leads to

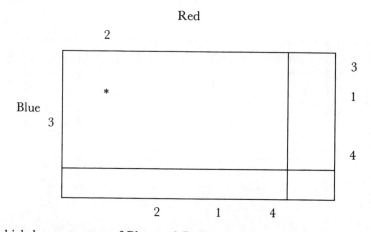

which has a new set of Blue and Red names adjacent to the border. Therefore we will find a new basic solution if we complete the schema. The result is

SCHEMA 5

(SOLUTION 2. BLUE 17:7:0:12, RED 0:0:1:3)

Red

	2					
	432	783	153	612	324	3
	$-8,640$	-720	720	0	0	1
Blue						
3	414,720	0	$-34,560*$	8,640	0	
	35,856	189	-261	1,044	972	4
	1,728	252	612	432	1,296	$D = -25,920$
	2	1	4			

The strategies are Blue 612:252:0:432 and Red 0:0:324:972, or, removing the common factors, Blue 17:7:0:12 and Red 0:0:1:3.

We now seek a negative pivot in the third row of Schema 5. There is only one possibility, namely $P = -34,560$, in the third column. If we pivot there, the set of strategies adjacent to the border becomes

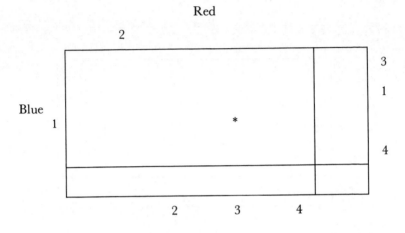

Red

which is unlike the sets in Solutions 1 and 2; so we will find another basic solution. It is given in Schema 6.

Schema 6

(Solution 3. blue 0:28:51:65, red 0:0:1:3)

Red

	2					
	3,024	1,044	153	−765	432	3
	0	−960*	720	240	0	1
Blue 1	−414,720	0	−25,920	−8,640	0	
	43,632	252	−261	1,305	1,296	4
	12,096	336	612	780	1,728	$D = -34,560$
	2	3	4			

We next return to row 2 and again seek a negative pivot. $P = -960$ is the sole possibility. It produces the following set of strategies adjacent to the border:

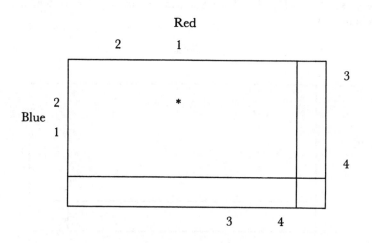

Red

This is another new set; so we continue to the next schema and solution:

Schema 7

(Solution 4. blue 0:0:1:1, red 0:0:1:3)

Red

	2	1				
	84	1,044	26	−14	12	3
2	0	−34,560	−720	−240	0	
1	−11,520*	0	−720	−240	0	
	1,212	252	−2	38	36	4
	336	336	24	24	48	$D = -960$

Blue (rows 2 and 1)

3   4

We go now to row 3 again and test the potential pivots by calculating the several values of c/p. These are $336/(-11,520) = -.029$, $24/(-720) = -.033$, and $24/(-240) = -.1$. The first is closest to zero; so $P = -11,520$. This pivot changes the border strategy set of Schema 7 to

Red

1

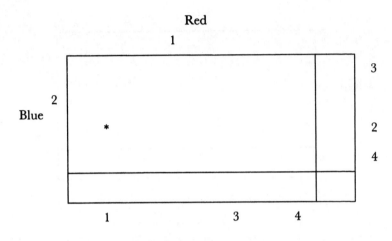

Blue   2

        3

        2

        4

1   3   4

which differs from those of the previous solutions; so we press on to

Schema 8

(Solution 5. blue 28:0:3:17, red 0:0:1:3)

Red

1

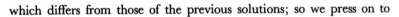

	84	12,528	249	−189	144	3
2	0	−414,720	−8,640*	−2,880	0	
Blue	−960	0	720	240	0	2
	1,212	3,024	−933	153	432	4
	336	4,032	36	204	576	$D = -11,520$

1   3   4

Turn again to row 2—for the last time in this sequence. The values of c/p for the three potential pivots are $4{,}032/(-414{,}720) = -.010$, $36/(-8{,}640) = -.004$, and $204/(-2{,}880) = -.07$; so $P = -8{,}640$ is the pivot.

This leads to the following set of border strategies:

Red

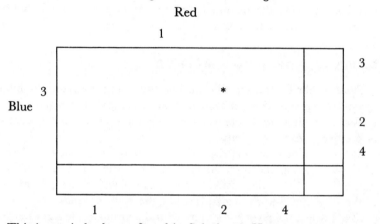

This is precisely the set found in Solution 1. Hence our sequence of pivots, alternating between row 2 and row 3, has led us from Solution 1 to Solution 2, then to Solutions 3, 4, and 5, and finally back to Solution 1.

We must now start again with Solution 1 and try a sequence of pivots beginning with row 3. The sole possibility is $P = -11{,}520$, and it leads to this border strategy set:

Red

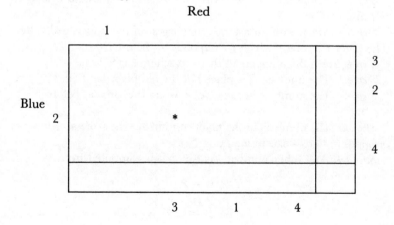

But this is the same set as in Solution 5; so we need not continue.

We have exhausted the possibilities—and ourselves as well—and have found five basic solutions. It is clear from this example that the task of finding solutions to games by the pivot method is simple and straightforward, but that the quantity of arithmetic required is substantial. Therefore the arithmetic deserves great care; so it will not be necessary to redo it. It suggests the concluding section of this chapter.

SUMMARY OF PIVOT METHOD

Step 1. Add a constant to all elements of the game matrix to eliminate negative elements. Also, if it is feasible, multiply all elements by a constant to eliminate fractions. It is desirable that the matrix consist of positive integers and zeros.

Step 2. Augment the matrix with a border of -1's along the lower edge, $+1$'s along the right edge, 0 in the corner, and an auxiliary number $D = 1$ near the corner. The entire display, including the names of the Blue and Red strategies, is called the First Schema.

Step 3. Select the pivot P—a number and position in the game matrix section of the schema.

Step 3.1. The potential pivot p must be positive.

Step 3.2. The border number below p—call it c—must be negative.

Step 3.3. The border number to the right of p is called r. Compute the criterion $-\dfrac{r \times c}{p}$ for every p.

Step 3.4. Underline the value of the criterion closest to 0 in each schema.

Step 3.5. Mark with an asterisk the largest of the underlined values. The pivot P is that p which corresponds to the asterisk.

Step 4. Form the numbers in the next schema as follows:

Step 4.1. The number D is placed in the old pivot position.

Step 4.2. The numbers in the pivot row are the same as before except for D.

Step 4.3. The numbers in the pivot column are the same as before but of opposite signs—except for D.

Step 4.4. Any other number—call it N—is computed from

$$\frac{N \times P - R \times C}{D}$$

where R and C are the numbers which share rows and columns with P and N.

Step 4.5. The next value of D is P.

Step 5. Complete the next schema:

Step 5.1. Exchange the Blue strategy name that is at the left of the pivot row with the name that is below the pivot column.

Step 5.2. Exchange the Red strategy name that is above the pivot column with the name that is at the right of the pivot row.

Step 6. Check the border of the new schema for negative numbers.

Step 6.1. If it contains one or more negative numbers, go again to *Step 3* and do *Steps 3, 4, 5,* and *6.*

Step 6.2. If it contains no negative numbers, we have found a basic solution to the game: The oddment for each active strategy is the border number adjacent to a strategy name, and the value of the game is the ratio D divided by the corner border number.

Step 7. If the border contains one or more 0's, find other basic solutions by pivoting, perhaps repeatedly, using all possible sequences of rows and columns which have border zeros. To find the pivots:

Step 7.1. If the border element of a column is 0, choose as pivot P that positive element p for which the ratio r/p is smallest.

Step 7.2. If the border element of a row is 0, choose as pivot P that negative element p for which the ratio c/p is closest to zero.

Step 8. Transfer strategy names about the pivot, according to *Step 5.*

Step 8.1. If the resulting border set of Blue and Red names is novel, calculate a new schema and a new basic solution by returning to *Step 4,* but modify the rules as follows if the pivot P is negative: Make the old value of D negative and change the signs of the elements discussed in *Steps 4.2* and *4.3.*

Step 8.2. If the resulting border set of Blue and Red names is not novel, the basic solutions that can be reached by the present sequence of 0's have been found.

HOW TO CHECK THE WORK

It is an unfortunate fact that any error made during the solution process is likely to lead to a nonsense answer, but to an answer that seems respectable. Therefore, it is important that the work be verified.

An absolute verification is available at the end: Try Blue's presumed-optimal strategy against each of Red's pure strategies. If Blue's strategy is indeed optimal, he will achieve precisely the value of the game against the strategies that are active in Red's optimal strategy, and at least that value against other Red strategies. Similarly, try Red's presumed-optimal strategy against Blue's pure strategies.

In the first example of this chapter, we found the solution Blue $0:1:5$, Red $1:0:2$ for the game on page 219. There Blue $0:1:5$ yields

$$\frac{6 \times 0 + 8 \times 1 + 4 \times 5}{0 + 1 + 5} = \frac{28}{6}$$

against Red 1;

$$\frac{0 \times 0 + (-2) \times 1 + 6 \times 5}{0 + 1 + 5} = \frac{28}{6}$$

against Red 2; and

$$\frac{3 \times 0 + 3 \times 1 + 5 \times 5}{0 + 1 + 5} = \frac{28}{6}$$

against Red 3; i.e., Blue always achieves the value of the game, $4\frac{2}{3}$; so his strategy is optimal.

Again, Red $1:0:2$ yields

$$\frac{6 \times 1 + 0 \times 0 + 3 \times 2}{1 + 0 + 2} = \frac{12}{3}$$

against Blue 1;

$$\frac{8 \times 1 + (-2) \times 0 + 3 \times 2}{1 + 0 + 2} = \frac{14}{3}$$

against Blue 2; and

$$\frac{4 \times 1 + 6 \times 0 + 5 \times 2}{1 + 0 + 2} = \frac{14}{3}$$

against Blue 3; i.e., Red achieves the value of the game, $4\frac{2}{3}$, against Blue's active strategies, and more otherwise—recalling that Red, the minimizing player, prefers the $\frac{12}{3}$ payoff to the $\frac{14}{3}$ value of the game.

We have verified the solution in this example. But suppose it had not checked out? It would be unpleasant to have to begin over and possibly to make the same mistakes again.

Fortunately there are built-in signals which often indicate the presence of errors. Perhaps the best of these is

If the game matrix consists entirely of integers, all schemata will consist entirely of integers.

That is self-evident as regards the numbers in the pivot row and pivot column because they are copies—except for signs in some cases—of numbers in the preceding schema; but it is not immediately evident that the other numbers, computed from the formula

$$\frac{N \times P - R \times C}{D}$$

will always be integers. But they will be, except possibly when an error is present.

There is another partial check available in every schema:

The sum of the oddments, for the Blue strategies whose names are listed along the lower border, is equal to the border number in the corner.

Similarly,

The sum of the oddments, for the Red strategies whose names are listed along the right border, is equal to the border number in the corner.

Another danger signal is the appearance of a negative number in the right border:

The numbers in the right border are never negative.

They may be zero, however. A negative number may indicate that the proper pivot was not selected.

Another danger signal may be provided by the ratios of the auxiliary numbers D to the border numbers in the corners of the schemata:

These ratios should not increase in value as we go from schema to schema.

They should decrease or remain constant, always. In the final schema, the ratio is equal to the value of the game; the sequence of ratios approaches that value from above and never recedes from it.

The checks mentioned above are adequate for most small games and for most game solvers. However, a large game may warrant an additional investment in checking machinery, such as the use of control sums.

CONTROL SUMS

To introduce control sums one adds another row, or another column, or both, to the First Schema; we shall do both, adjoining them to the upper and left edges of the schema. We demonstrate, using the first example of this chapter:

FIRST SCHEMA

Red

	1	2	3		
1	8	2	5	1	
2	10	0	5	1	
3	6	8	7*	1	
	-1	-1	-1	0	$D = 1$

The number—call it n—in the top cell of the new column is chosen so as to make the sum of the numbers in the first row equal to -1; e.g.,

$$n + 8 + 2 + 5 + 1 = -1$$
or
$$n = -17$$

Similarly, the second number n in the new column is chosen so as to make the sum of the second row numbers equal to -1; i.e.,

$$n + 10 + 0 + 5 + 1 = -1$$
or
$$n = -17$$

again—a coincidence. The last two column numbers are -23 and 2.

The numbers n in the new row are chosen so as to make the sums of the column numbers equal to $+1$—rather than -1; e.g., the first number is

$$n + 8 + 10 + 6 - 1 = 1$$

or $$n = -22$$

and the others are -8, -15, and -2. The First Schema then becomes

FIRST SCHEMA

Red

		1	2	3	
		−22	−8	−15	−2
1	−17	8	2	5	1
Blue 2	−17	10	0	5	1
3	−23	6	8	7*	1
	2	−1	−1	−1	0

$D = 1$

The significant fact about this arrangement is that

In every schema, the sums of the numbers in the long rows and columns will be equal, respectively, to the numbers $-D$ *and* D.

Decisions regarding pivots are made as if the control sums were not present. However, they are treated as ordinary schema elements when the next schema is calculated. Thus, by the rules of *Steps 4* and *5*, we go progressively to the Second Schema:

Red

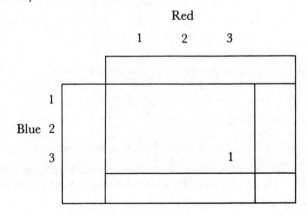

1 2 3

Blue

1

2

3 1

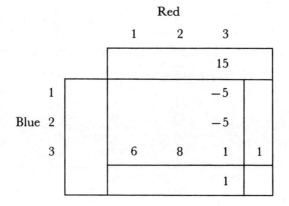

Finally,

SECOND SCHEMA

Red

		1	2			
		−64	64	15	1	
Blue	1	−4	26	−26	−5	2
	2	−4	40*	−40	−5	2
		−23	6	8	1	1	3
		−9	−1	1	1	1	D = 7

3

The sums are all as expected; e.g., the second row is

$$-4 + 40 - 40 - 5 + 2 = -7$$

and the second column is

$$64 - 26 - 40 + 8 + 1 = 7$$

i.e., $-D$ and D, respectively; so we may proceed with the next pivot operation. The next pivot, 40, was determined earlier and has been marked by an asterisk.

The Third Schema is

Red

2

		64	0	40	24	
1	− 8	− 26	0	− 10	4	
	− 4	7	− 40	− 5	2	1
Blue	− 128	− 6	80	10	4	3
	− 52	1	0	5	6	$D = 40$

2   3

Again the row and columns are correct—equal to −40 and to 40, respectively—so the work is probably correct.

No matter what system of checks is used, there is no substitute for trying the apparent solution in the original game. In an ultimate effort to convince the reader of this, the author now resorts to confession and to the vertical pronoun: I may, while writing this book, have solved more games than any other person. This effort surely entitles me to cheat occasionally, as by omitting the absolute check. However, it turns out that I am almost always attended on such occasions by an evil familiar—doubtless the fellow who originated the above argument about my prerogatives—who introduces unbelievable stupidities in my work. If everything else is perfect, it will turn out that I have not copied the matrix correctly. If I enlist the aid of a faultless modern computer, my confidence reaches new heights—and my fall is greater. I hate to subscribe to a demon theory of errors, but there it is.

Table of Random Digits

11	16	43	63	18	75	06	13	76	74	40	60	31	61	52	83	23	53	73	61
21	21	59	17	91	76	83	15	86	78	40	94	15	35	85	69	95	86	09	16
10	43	84	44	82	66	55	83	76	49	73	50	58	34	72	55	95	31	79	57
36	79	22	62	36	33	26	66	65	83	39	41	21	60	13	11	44	28	93	20
73	94	40	47	73	12	03	25	14	14	57	99	47	67	48	54	62	74	85	11
49	56	31	28	72	14	06	39	31	04	61	83	45	91	99	15	46	98	22	85
64	20	84	82	37	41	70	17	31	17	91	40	27	72	27	79	51	62	10	07
51	48	67	28	75	38	60	52	93	41	58	29	98	38	80	20	12	51	07	94
99	75	62	63	60	64	51	61	79	71	40	68	49	99	48	33	88	07	64	13
71	32	55	52	17	13	01	57	29	07	75	97	86	42	98	08	07	46	20	55
65	28	59	71	98	12	13	85	30	10	34	55	63	98	61	88	26	77	60	68
17	26	45	73	27	38	22	42	93	01	65	99	05	70	48	25	06	77	75	71
95	63	99	97	54	31	19	99	25	58	16	38	11	50	69	25	41	68	78	75
61	55	57	64	04	86	21	01	18	08	52	45	88	88	80	78	35	26	79	13
78	13	79	87	68	04	68	98	71	30	33	00	78	56	07	92	00	84	48	97
62	49	09	92	15	84	98	72	87	59	38	71	23	15	12	08	58	86	14	90
24	21	66	34	44	21	28	30	70	44	58	72	20	36	78	19	18	66	96	02
16	97	59	54	28	33	22	65	59	03	26	18	86	94	97	51	35	14	77	99
59	13	83	95	42	71	16	85	76	09	12	89	35	40	48	07	25	58	61	49
29	47	85	96	52	50	41	43	19	66	33	18	68	13	46	85	09	53	72	82
96	15	59	50	09	27	42	97	29	18	79	89	32	94	48	88	39	25	42	11
29	62	16	65	83	62	96	61	24	68	48	44	91	51	02	44	12	61	94	38
12	63	97	52	91	71	02	01	72	65	94	20	50	42	59	68	98	35	05	61
14	54	43	71	34	54	71	40	24	01	38	64	80	92	78	81	31	37	74	00
83	40	38	88	27	09	83	41	13	33	04	29	24	60	28	75	66	62	69	54
67	64	20	52	04	30	69	74	48	06	17	02	64	97	37	85	87	51	21	39
64	04	19	90	11	61	04	02	73	09	48	07	07	68	48	02	53	19	77	37
17	04	89	45	23	97	44	45	99	04	30	15	99	54	50	83	77	84	61	15
93	03	98	94	16	52	79	51	06	31	12	14	89	22	31	31	36	16	06	50
82	24	43	43	92	96	60	71	72	20	73	83	87	70	67	24	86	39	75	76
96	99	05	52	44	70	69	32	52	55	73	54	74	37	59	95	63	23	95	55
09	11	97	48	03	97	30	38	87	01	07	27	79	32	17	79	42	12	17	69
57	66	64	12	04	47	58	97	83	64	65	12	84	83	34	07	49	32	80	98
46	49	26	15	94	26	72	95	82	72	38	71	66	13	80	60	21	20	50	99
08	43	31	91	72	08	32	02	08	39	31	92	17	64	58	73	72	00	86	57
10	01	17	50	04	86	05	44	11	90	57	23	82	74	64	61	48	75	23	29
92	42	06	54	31	16	53	00	55	47	24	21	94	10	90	08	53	16	15	78
35	54	25	58	65	07	30	44	70	10	31	30	94	93	87	02	33	00	24	76
86	59	52	62	47	18	55	22	94	91	20	75	09	70	24	72	61	96	66	28
72	11	53	49	85	58	03	69	91	37	28	53	78	43	95	26	65	43	78	51

```
07 42 85 88 63    96 02 38 89 36    97 92 94 12 20    86 43 19 44 85
35 37 92 79 22    28 90 65 50 13    40 56 83 32 22    40 48 69 11 22
10 98 22 28 07    10 92 02 62 99    41 48 39 29 35    17 06 17 82 52
90 12 73 33 41    77 80 61 24 46    93 04 06 64 76    24 99 04 10 99
63 00 21 29 90    23 51 06 87 74    76 86 93 93 00    84 97 80 75 04
40 77 98 63 82    48 45 46 52 69    02 98 25 79 91    50 76 59 19 30
43 21 61 26 08    18 16 78 46 31    94 47 97 65 00    39 17 00 66 29
96 16 76 43 75    74 10 89 36 43    52 29 17 58 22    95 96 69 09 47
70 97 56 26 93    35 68 47 26 07    03 68 40 36 00    52 83 15 53 81
85 81 26 18 75    23 57 07 57 54    58 93 92 83 66    86 76 56 74 65

37 10 06 24 92    63 64 24 76 38    54 72 35 65 27    53 07 63 82 35
53 40 61 38 55    38 51 92 95 00    84 82 88 12 48    25 54 83 40 75
55 17 28 15 56    18 85 65 90 43    65 79 90 19 14    81 36 30 51 73
40 35 38 48 07    47 76 74 68 90    87 91 73 85 49    48 21 37 17 08
18 89 90 96 12    77 54 15 76 75    26 90 78 81 73    71 18 92 83 77
68 14 12 53 40    92 55 11 13 26    68 05 26 54 22    88 46 00 63 52
51 55 99 11 59    81 31 06 32 51    42 58 76 81 49    88 14 79 97 00
92 21 43 33 86    73 45 97 93 59    97 17 65 54 16    67 64 20 50 51
15 08 95 05 57    33 16 68 70 94    53 29 58 71 33    38 26 49 47 08
96 46 10 06 04    11 12 02 22 54    23 01 19 41 08    29 19 66 51 87

28 17 74 41 11    15 70 57 38 35    75 76 84 95 49    24 54 36 32 85
66 95 34 47 37    81 12 70 74 93    86 66 87 03 41    66 46 07 56 48
19 71 22 72 63    84 57 54 98 20    56 72 77 20 36    50 34 73 35 21
68 75 66 47 57    19 98 79 22 22    27 93 67 80 10    09 61 70 44 08
75 02 26 53 32    98 60 62 94 51    31 99 46 90 72    37 35 49 30 25
11 32 37 00 69    90 26 98 92 66    02 98 59 53 03    15 18 25 01 66
55 20 86 34 70    18 15 82 52 83    89 96 51 02 06    95 83 09 54 06
11 47 40 87 86    05 59 46 70 45    45 58 72 96 11    98 57 94 24 81
81 42 28 68 42    60 99 77 96 69    01 07 10 85 30    74 30 57 75 09
21 77 17 59 63    23 15 19 02 74    90 20 96 85 21    14 29 33 91 94

42 27 81 21 60    32 57 61 42 78    04 98 26 84 70    27 87 51 54 80
17 69 76 01 14    63 24 73 20 96    19 74 02 46 37    97 37 73 21 12
05 68 63 02 43    34 12 40 29 36    50 19 77 98 69    86 49 76 87 09
52 99 24 66 50    89 91 05 73 95    46 95 46 75 36    28 96 88 19 36
94 51 89 39 84    81 47 86 77 50    82 54 96 26 76    31 12 34 98 99
00 18 47 21 86    78 90 67 54 80    61 79 88 16 00    80 01 88 47 42
87 46 26 31 65    79 81 66 16 30    57 66 62 90 55    46 51 80 14 87
88 69 25 87 16    12 27 34 81 76    29 80 56 49 94    66 87 26 22 30
20 09 44 29 62    41 38 21 67 68    06 71 13 49 39    19 59 97 62 47
60 93 58 15 04    50 52 08 21 53    13 93 44 68 85    58 31 58 83 66
```

51 39 28 59 36	43 89 85 05 96	28 54 99 83 27	99 94 32 53 77
54 23 94 19 18	79 52 64 62 74	40 87 16 18 03	25 76 75 54 84
57 89 27 33 94	07 16 09 02 62	47 70 43 83 55	71 70 88 01 17
02 33 07 47 36	53 27 44 44 68	62 61 11 96 98	09 30 42 92 65
76 11 52 92 47	55 34 25 12 99	03 04 78 39 81	11 91 60 92 67
63 31 28 18 86	29 08 52 01 01	26 46 05 05 01	31 73 11 89 38
27 63 22 15 70	34 27 45 64 26	01 76 42 59 59	69 29 38 98 75
06 33 56 21 11	44 01 45 25 67	11 76 25 48 06	02 65 15 29 12
64 14 28 76 76	21 35 88 87 73	31 73 63 16 95	11 52 36 42 13
28 43 62 54 68	75 23 57 53 70	97 15 54 87 06	52 23 92 18 31
09 52 28 38 55	85 97 31 58 88	31 18 14 96 72	17 23 70 40 24
93 71 41 54 14	93 71 20 27 42	32 11 58 26 83	67 18 28 90 30
15 68 15 35 99	58 18 57 38 40	07 06 87 59 47	71 74 36 92 85
77 71 22 39 14	08 90 74 37 68	26 62 27 41 84	75 16 69 67 48
78 45 35 48 44	61 50 90 12 45	02 80 55 26 76	22 51 94 78 48
24 86 06 82 84	19 36 72 90 73	32 30 15 87 01	04 19 33 01 42
37 28 40 68 44	78 88 75 72 76	26 33 95 69 09	39 33 14 21 01
35 48 85 24 73	37 63 43 25 69	95 27 40 95 08	81 01 24 24 13
51 59 55 99 09	35 22 34 49 91	24 27 53 96 32	09 77 79 88 00
90 66 03 51 71	30 02 19 11 20	36 11 64 21 28	65 40 19 41 99
47 50 50 20 08	20 30 08 71 88	96 19 50 70 59	13 26 63 13 89
13 35 00 84 14	64 04 99 43 77	22 40 89 49 58	19 09 55 80 35
33 00 69 26 90	69 24 89 74 43	53 89 62 35 08	16 22 75 69 29
55 21 66 38 86	06 80 41 18 61	22 56 50 24 75	00 25 87 90 18
21 99 12 62 28	14 80 11 91 92	49 43 82 07 72	60 84 66 97 32
71 02 52 82 12	10 47 42 75 22	65 62 03 46 84	00 21 00 48 63
65 52 21 52 42	84 55 47 45 60	20 24 62 69 41	41 29 80 47 63
27 97 55 49 23	90 65 00 61 70	09 43 30 91 67	35 16 63 27 31
07 30 00 97 04	36 09 96 15 77	95 55 27 34 56	16 57 88 81 40
54 35 71 36 89	19 56 90 38 14	76 05 30 51 50	69 12 56 94 42
00 97 70 44 81	42 04 40 86 49	34 82 23 58 43	78 46 88 23 80
13 92 07 87 61	12 31 19 28 08	07 75 30 40 73	58 52 08 00 22
08 39 53 70 43	37 88 03 41 72	04 20 49 44 34	62 79 88 19 02
46 16 66 72 06	01 61 94 37 69	96 77 01 94 40	29 70 04 20 93
87 76 77 76 07	03 74 20 16 13	65 98 96 28 43	10 91 73 44 58
29 88 09 52 88	21 64 44 65 87	06 64 49 47 84	66 99 56 18 12
36 24 83 66 66	14 89 45 92 73	88 95 04 60 77	34 65 11 20 38
12 38 62 96 56	30 47 42 59 64	21 48 29 54 22	02 00 23 36 71
52 06 87 38 01	52 18 81 94 91	55 13 76 10 39	02 00 66 99 13
41 72 75 21 71	56 71 90 60 54	98 44 18 15 29	59 60 76 52 25

```
49 31 97 45 80     57 47 01 46 00     57 16 83 04 58     23 89 20 78 25
88 78 67 69 63     12 12 72 50 14     71 88 66 53 34     38 01 30 93 79
84 86 69 52 02     43 98 37 26 55     40 41 85 95 04     52 38 30 72 32
11 84 92 64 82     20 46 19 94 50     28 83 37 66 61     47 27 79 29 35
54 96 61 75 94     57 39 37 32 67     37 88 36 21 24     62 19 94 95 42
10 95 93 33 49     80 71 99 67 51     44 88 23 35 92     66 23 41 38 21
22 78 40 77 83     35 90 30 00 91     19 08 21 38 73     07 18 42 15 66
86 03 76 17 91     33 81 56 39 68     45 31 62 92 83     89 31 85 58 06
80 03 76 50 89     85 91 97 43 91     22 78 85 54 33     31 18 87 48 82
72 75 18 43 59     15 76 91 36 15     08 29 38 61 93     05 02 62 12 55

18 53 20 38 74     66 22 07 90 50     29 22 37 05 41     67 11 58 45 84
22 93 62 20 58     49 17 11 10 27     22 68 18 01 10     31 59 50 92 46
66 39 77 65 10     81 15 00 07 04     74 58 09 03 54     43 74 42 21 78
89 73 02 32 72     65 42 03 50 91     69 09 37 13 64     08 10 79 69 52
81 82 17 53 23     96 06 89 17 24     40 45 69 12 34     58 09 06 53 42
94 37 78 25 54     53 58 61 14 32     72 92 76 73 49     83 96 25 89 12
68 48 54 99 91     53 16 51 98 65     61 86 93 30 93     81 12 90 64 81
07 33 00 71 84     86 78 86 45 77     40 04 81 65 20     07 63 81 07 97
10 99 31 49 30     35 07 23 64 29     68 77 39 76 69     28 65 68 99 38
20 80 11 51 78     64 45 38 33 57     09 77 43 07 51     49 74 01 13 85

79 24 13 53 47     66 85 17 92 47     46 13 93 66 89     82 58 71 35 86
43 59 33 95 55     97 34 55 84 94     26 56 69 53 23     32 99 38 99 88
29 52 26 27 13     33 70 11 71 86     06 76 55 71 41     48 61 71 82 82
88 83 64 72 90     67 27 47 83 62     35 38 49 03 80     12 31 78 97 02
65 90 56 62 53     91 48 23 06 89     49 33 37 84 82     36 19 91 13 55
44 79 86 93 71     07 86 59 17 56     45 59 51 40 44     56 80 69 91 26
35 51 09 91 39     32 03 12 79 25     79 81 91 50 54     76 17 41 22 06
50 12 59 32 23     64 20 94 97 14     11 97 16 22 34     74 85 74 64 01
25 17 39 00 38     63 87 14 04 18     11 45 28 93 18     53 08 42 19 93
68 45 99 00 94     44 99 59 37 18     38 74 68 12 71     96 26 09 81 37

93 36 91 30 44     69 68 67 81 62     66 37 80 29 19     34 01 25 00 80
19 36 05 50 49     94 95 17 63 41     84 01 93 06 90     25 65 67 29 96
47 79 88 98 90     06 89 36 54 83     17 70 12 12 92     14 88 01 53 86
69 22 33 20 07     03 51 36 11 49     32 54 69 20 72     62 52 22 15 04
34 51 15 07 21     84 85 03 41 59     97 13 86 19 19     97 78 92 85 75
54 03 15 93 29     58 96 35 22 20     35 29 22 79 24     55 46 74 30 36
66 72 28 55 15     04 72 39 24 11     02 73 70 81 68     30 04 36 34 50
71 05 90 74 96     38 40 41 81 26     28 26 13 78 44     12 54 31 43 98
45 47 88 60 66     31 13 53 32 43     80 57 33 06 06     48 64 45 30 08
97 24 69 11 21     89 43 72 03 93     77 15 38 85 52     26 84 31 28 44
```

22	98	22	59	36	96	41	73	48	45	85	14	95	75	04	15	05	93	68	49
48	24	36	29	93	47	13	28	52	48	35	22	97	28	37	36	75	27	16	55
93	51	41	49	15	67	96	08	22	03	40	11	72	43	46	32	18	98	70	74
69	70	79	83	03	93	06	91	62	16	60	87	59	75	45	68	65	29	21	60
87	46	79	17	94	70	81	41	27	43	03	76	93	25	51	74	80	14	16	92
81	00	68	14	98	59	37	53	05	02	94	07	79	22	09	31	50	66	96	06
15	45	88	34	81	50	18	74	33	75	94	37	60	06	66	94	14	52	23	99
33	46	91	25	10	23	09	54	80	16	42	35	41	13	47	90	92	00	38	64
67	19	80	71	76	65	99	61	83	17	81	14	94	32	91	10	81	74	43	48
58	03	79	22	61	85	50	45	56	90	10	63	17	82	38	00	15	74	62	59
84	98	36	83	12	25	51	95	61	58	86	30	00	76	89	14	00	67	77	53
35	55	40	29	35	72	88	96	87	72	19	85	03	96	50	65	22	21	55	63
04	36	81	76	32	50	96	27	19	08	94	46	46	64	32	62	24	31	36	74
81	31	16	04	79	69	98	53	09	52	23	92	14	97	30	21	71	89	23	14
03	82	38	98	87	55	82	87	44	52	72	77	52	37	16	42	85	37	47	93
80	42	26	54	37	38	79	75	62	61	27	81	64	67	04	82	73	50	33	39
61	30	74	94	68	43	34	44	37	00	20	20	77	70	88	17	16	72	45	31
83	87	38	25	57	10	00	28	00	93	59	28	30	44	94	60	72	52	14	31
38	11	01	68	55	28	92	29	37	58	88	73	13	63	76	51	38	35	76	19
43	89	29	11	89	87	22	65	69	35	84	76	26	79	96	75	00	00	17	45
93	68	30	96	64	53	92	74	98	85	20	75	49	23	55	57	95	51	09	40
32	74	80	21	21	11	97	29	69	14	28	06	56	95	64	06	83	55	68	45
49	21	19	29	63	38	62	56	53	12	62	17	57	33	53	84	97	21	77	26
63	36	56	42	24	69	47	55	75	12	11	04	45	04	83	68	82	19	74	26
63	57	62	63	73	44	61	04	37	48	00	33	16	34	22	99	62	27	67	57
41	07	84	70	36	65	52	46	84	66	67	15	72	64	19	37	97	81	65	11
70	84	68	95	58	64	17	31	53	81	87	71	35	08	41	46	27	02	65	08
68	80	06	44	92	20	16	23	27	07	10	28	18	25	25	74	15	58	67	49
44	97	78	95	25	51	26	96	37	47	91	36	77	40	33	67	02	06	90	92
79	35	46	38	47	24	39	55	36	79	40	56	03	69	14	69	17	63	19	18
14	95	42	22	99	40	15	65	26	85	29	22	33	83	83	30	31	57	09	99
01	71	19	84	39	09	44	63	39	37	49	09	54	02	38	81	69	71	24	74
62	32	85	53	28	45	73	89	39	40	27	46	62	69	27	53	34	51	13	79
73	00	46	21	09	81	90	77	10	77	57	46	37	00	45	65	12	34	90	70
34	21	88	94	45	05	60	95	23	36	50	55	89	22	42	52	73	28	15	02
99	15	90	19	68	45	88	68	68	75	28	41	39	59	18	44	15	64	69	59
92	85	82	99	49	15	81	79	33	72	56	65	74	31	93	58	13	05	42	73
27	39	69	74	77	65	55	47	16	01	13	12	16	88	67	95	76	35	96	67
37	10	34	53	09	30	12	94	33	80	96	99	68	93	56	22	78	46	01	84
57	34	79	70	12	48	42	82	06	06	60	74	22	22	26	89	99	32	45	97

29 41 72 09 72	84 13 42 91 66	71 99 34 28 19	28 79 91 30 95
70 18 50 54 60	07 35 38 58 55	11 83 80 22 91	89 88 24 16 13
06 39 35 20 17	21 23 62 05 30	14 30 05 53 19	47 90 83 92 32
42 60 59 24 74	09 36 14 24 04	33 72 24 77 22	12 54 72 97 59
26 48 34 88 73	84 26 03 32 85	51 27 58 48 21	87 44 30 24 91
93 60 97 74 25	57 26 86 93 20	13 56 80 08 25	80 83 58 36 39
19 79 00 32 65	68 47 42 96 78	08 45 60 14 75	40 31 77 20 77
63 43 46 72 51	20 51 08 06 27	62 61 86 40 27	64 47 75 04 32
76 63 64 27 71	12 58 35 63 29	59 51 22 52 49	42 18 50 96 22
51 25 87 43 30	47 75 15 26 72	10 99 59 19 22	50 21 71 41 87
13 26 01 42 38	88 39 15 08 58	60 73 55 89 65	58 49 20 96 66
31 81 15 09 35	89 51 01 55 66	22 09 62 63 38	30 24 00 43 43
48 83 26 95 24	32 11 34 66 73	24 96 59 71 07	41 85 68 52 96
32 36 71 84 57	03 31 22 55 64	75 10 99 31 42	91 95 16 88 94
90 34 61 74 76	71 03 92 14 97	64 00 15 42 64	71 66 79 42 62
98 51 30 86 79	69 41 38 20 67	68 25 07 45 01	48 84 60 91 00
08 48 41 56 17	82 79 00 83 24	95 09 12 70 82	73 94 95 95 33
60 04 28 75 62	49 72 82 64 00	10 11 21 06 30	82 87 33 93 12
87 65 38 64 09	09 27 73 83 79	09 93 45 75 87	44 00 52 10 90
35 71 10 03 98	30 49 45 09 23	33 30 85 14 41	13 80 43 26 09
46 48 33 27 22	74 79 96 02 77	09 29 94 26 17	94 65 41 13 15
60 85 80 63 37	20 22 03 49 52	98 21 86 97 10	03 42 79 92 55
70 78 30 14 52	87 66 12 40 28	41 83 04 04 72	46 04 10 89 52
65 34 64 30 42	62 13 43 22 85	35 20 18 56 72	66 68 85 39 48
96 34 95 21 64	23 29 87 22 17	41 64 06 30 18	74 28 09 16 70
63 46 91 34 59	11 09 19 66 46	37 84 93 43 89	55 97 49 94 43
60 57 97 54 98	97 81 58 61 51	60 23 67 57 05	83 33 98 49 23
31 01 87 40 81	01 49 40 61 65	70 80 28 81 89	11 59 41 34 47
00 04 73 38 02	73 34 25 43 38	43 35 49 39 50	54 76 53 75 53
55 75 48 69 00	87 93 10 12 38	15 44 92 88 47	82 63 56 85 43
38 05 83 03 46	35 72 27 88 49	92 82 54 05 36	76 78 04 42 64
06 90 44 32 00	89 09 23 09 74	00 70 16 65 89	04 62 25 13 54
48 43 07 91 89	59 69 09 38 30	59 93 99 51 37	65 79 94 11 47
78 55 65 87 76	35 37 02 47 74	57 49 08 58 05	04 16 80 74 03
07 66 20 77 12	70 41 09 94 28	76 51 36 93 39	33 70 13 21 84
98 74 51 23 38	11 08 80 54 54	06 87 15 01 32	50 01 02 40 89
52 92 38 13 43	14 90 82 38 96	01 52 90 94 99	57 66 00 23 09
35 16 59 31 08	32 06 82 15 47	54 73 97 37 96	14 28 37 69 59
65 22 11 22 07	66 07 95 94 34	94 94 61 72 41	82 75 41 80 73
74 45 35 59 75	85 46 39 45 67	49 80 03 00 54	85 78 02 30 94

03	49	66	06	56	69	73	55	50	53	15	39	38	25	29	15	76	08	97	88
64	25	28	09	87	44	41	63	14	38	26	05	53	00	56	15	22	61	10	81
49	22	51	16	47	69	73	37	99	39	86	30	53	50	43	66	21	97	48	31
62	47	61	93	65	16	56	86	25	09	89	14	62	53	24	22	75	93	56	01
21	11	93	11	66	78	38	51	48	75	86	85	79	94	04	91	54	77	49	34
35	08	65	00	39	71	25	14	57	69	27	90	84	89	44	09	01	75	43	97
33	33	07	87	01	88	53	64	13	98	86	41	32	07	09	41	58	96	36	41
68	34	64	53	55	37	07	24	08	32	33	91	55	38	90	19	88	73	41	08
55	52	63	17	47	57	30	90	92	38	78	74	26	99	71	30	14	87	26	47
86	02	05	57	48	08	63	66	37	73	07	22	80	50	82	72	82	19	10	26

Solutions to Exercises

Solutions to the Exercises are given here. In comparing your results with these, the following two remarks are important:

Remark 1. All oddments are expressed in their simplest forms; i.e., factors common to all oddments of a set have been eliminated. Thus, the odds 8:24:16 would be written here as 1:3:2.

Remark 2. While this point is not brought out early in the text, the fact is that many games have more than one basic solution. We have tried to provide all basic solutions for the games in the first several sets of Exercises. The alternative solutions are lettered (*a*), (*b*), etc. Note, however, that you may find a valid solution which does not appear in this list; if so, it will be a combination of two or more of these.

In the later sets of Exercises, for 4×4 games and larger ones, we have not attempted to provide all solutions. However, by the time you are ready to solve these larger games, you should have no difficulty in establishing that your results are correct.

The values of the games are unique; so your results should always agree with ours on that point.

EXERCISES 1

Exercise	Blue 1	Blue 2	Red 1	Red 2	Value of Game
1	1	0	0	1	4
2	0	1	1	0	6
3	0	1	0	1	−7
4	0	1	0	1	6
5	0	1	0	1	3
6	0	1	1	0	3
7	1	1	1	3	$\frac{5}{2}$
8	1	1	1	3	$\frac{3}{2}$
9	1	1	1	3	$\frac{1}{2}$
10	1	1	1	3	0
11	1	1	1	1	0
12	1	1	1	1	$-\frac{1}{2}$
13	1	1	1	1	$\frac{9}{2}$
14	1	1	1	1	$9\frac{3}{2}$
15	1	1	5	4	0
16	8	9	10	7	$\frac{5}{17}$
17	0	1	0	1	5
18	3	1	1	1	−1
19	10	1	10	111	$10\frac{10}{11}$
20	4	3	1	1	0

EXERCISES 2

Exercise	Blue 1	Blue 2	Red 1	Red 2	Red 3	Red 4	Red 5	Red 6	Red 7	Red 8	Value of Game
1	1	0	0	1	0						0
2	1	0	0	1	0	0					-6
3	3	1	0	1	7						$7/4$
4	7	9	3	0	0	0	0	13	0	0	$29\frac{3}{16}$
5	1	1	0	1	1	0					$5/2$
6	10	11	0	0	1	20					$11\frac{1}{21}$
7	5	7	7	5	0	0	0				$2\frac{5}{12}$
8	5	4	5	0	0	1	0				$-1/3$
9	7	6	0	19	0	0	7				$-3/13$
10	5	3	0	7	0	9	0				$-1\frac{3}{8}$
11	5	7	0	1	0	1					$7/2$
12	7	6	0	7	6						$4\frac{2}{13}$
13 (a)	1	1	4	5	0	0	0	0			5
(b)			3	0	5	0	0	0			
(c)			2	0	0	5	0	0			
(d)			0	1	0	0	4	0			
(e)			0	0	1	0	3	0			
(f)			0	0	0	1	2	0			
(g)			0	3	0	0	0	4			
(h)			0	0	3	0	0	3			
(i)			0	0	0	3	0	2			

Exercise	Blue 1	Blue 2	Blue 3	Blue 4	Blue 5	Blue 6	Blue 7	Blue 8	Red 1	Red 2	Value of Game
14	0	0	1						1	0	9
15	1	0	0	0	1				1	1	3
16	1	0	0						0	1	-3
17	1	1	0						1	1	$1/2$
18	0	0	1						1	1	1
19	1	0	0	0	0	1			1	1	2
20	1	0	0	2	0				2	1	$1\frac{9}{3}$

EXERCISES 3

Exercise	Blue			Red			Value of Game
	1	2	3	1	2	3	
1(a)	0	0	1	1	0	0	4
(b)				1	0	1	
2	0	0	1	0	0	1	5
3	0	1	0	1	0	0	2
4(a)	0	1	0	0	0	1	4
(b)				1	0	2	
(c)				0	9	7	
5	1	1	1	1	1	1	$\frac{2}{3}$
6	1	1	1	1	1	1	0
7	2	3	2	2	3	2	$\frac{8}{7}$
8	1	2	4	1	2	4	$1\frac{5}{7}$
9	18	5	22	25	9	11	$14\frac{2}{45}$
10	21	19	16	16	31	9	$6\frac{7}{56}$
11(a)	1	0	1	1	0	1	1
(b)				0	1	0	
12(a)	1	0	1	1	0	1	3
(b)	0	1	0				
13(a)	0	1	1	9	21	4	$\frac{3}{2}$
(b)				1	1	0	
14	7	0	1	1	0	3	$\frac{3}{4}$
15	0	7	1	1	0	3	$\frac{3}{4}$
16	1	0	6	3	0	4	$3\frac{1}{7}$
17	0	1	2	1	0	1	2
18	2	5	0	0	6	1	$1\frac{2}{7}$
19(a)	0	5	2	3	4	0	$2\frac{2}{7}$
(b)				6	1	7	
20	1	1	0	3	1	0	$\frac{5}{2}$
21	36	11	10	10	25	22	$22\frac{6}{57}$
22	15	19	40	24	35	15	$29\frac{7}{74}$
23	1	1	1	1	1	1	$1\frac{0}{3}$
24	1	1	1	8	2	11	$50\frac{9}{3}$
25	3	0	4	6	0	1	$2\frac{4}{7}$
26	0	1	9	3	0	2	$2\frac{3}{5}$
27	4	5	0	8	0	1	$4\frac{0}{9}$
28	11	9	15	8	14	13	$10\frac{8}{35}$
29(a)	0	5	3	1	4	3	$2\frac{3}{8}$
(b)				5	0	3	
30	4	1	0	2	3	0	$1\frac{2}{5}$
31	1	1	1	1	1	1	3

EXERCISES 4

	Blue			Red						Value of
Exercise	1	2	3	1	2	3	4	5	6	Game
1	0	1	0	0	0	1	0			4
2	0	0	1	0	0	0	1			6
3	1	8	9	9	8	1	0			$59/18$
4	7	5	1	0	4	3	6			$15/13$
5(*a*)	1	1	0	1	1	0	0			½
(*b*)				0	1	1	0			
(*c*)				0	0	1	1			
6(*a*)	1	1	0	1	0	3	0			3/2
(*b*)				1	0	0	1			
(*c*)				0	1	1	0			
7	10	9	23	0	5	8	1			$55/14$
8(*a*)	0	1	1	0	0	1	0			3
(*b*)	0	1	3							
9(*a*)	3	3	2	17	0	3	12	0		$17/8$
(*b*)				11	0	9	0	12		
10(*a*)	1	1	1	1	1	1	0	0	0	$10/3$
(*b*)				0	0	0	1	1	1	
(*c*)				3	2	0	1	0	0	
(*d*)				4	4	0	0	1	0	
(*e*)				2	0	0	3	0	1	
(*f*)				0	1	7	0	0	1	
(*g*)				0	4	0	0	7	4	
(*h*)				0	0	4	1	0	1	

	Blue							Red			Value of
Exercise	1	2	3	4	5	6	7	1	2	3	Game
11(*a*)	1	0	0	0	1			1	0	2	0
(*b*)	0	1	0	1	0			0	2	1	
(*c*)	0	0	1	0	0						
(*d*)	0	2	0	0	1						
(*e*)	1	0	0	2	0						
12(*a*)	0	0	13	0	2	0		3	8	10	$-68/3$
(*b*)	13	0	0	0	17	0		4	6	11	
13	0	1	0	0	1	1	0	1	1	1	⅓
14	0	0	13	2				1	0	4	$17/5$
15(*a*)	0	0	0	1				1	0	0	3
(*b*)	1	0	0	1							

EXERCISES 5

Exercise	Blue								Red								Value of Game
	1	2	3	4	5	6	7	8	1	2	3	4	5	6	7	8	
1(a)	0	1	0	0					0	0	1	0					2
(b)	0	1	1	0					1	0	0						
2	0	1	0	0					0	0	0	1					5
3(a)	1	0	0	0					1	0	0	0					3
(b)									1	0	0	2					
4(a)	1	0	0	0					1	0	1	0					0
(b)									0	0	0	0	0				
(c)	0	0	0	1					0	0	1	0	1				
(d)									0	0	0	0	1				
5	8	3	7	9					5	7	3	3					2 3/9
6	76	92	78	89					61	72	161	41					68 6/35
7	109	130	11	276					132	235	68	91					210 1/26
8	0	0	1	2		0			0	1	0	0					5
9	38	0	0	21	12				0	0	8	14					147/71
10	61	72	161	41	29				76	49	0	92	78	89	0	0	68 6/35
11	28	33	31	21	378				4	0	6	57	40				149/142
12	520	247	233	348	19				332	35	34	266	245				670 3/726
13	5	51	9	25	405				1096	849	493	139	764				32 4/109
14	415	540	660	245	4				273	2086	535	665	440				276 5/453
15	0	0	1	0	0				1	1	0	0	0				5
16	0	0	0	5	4				0	2	0	1	0				9/8
17	2	0	1	0	0				0	0	0	5	4				1 1/3
18	36	11	0	0	10				10	0	0	22	0	1			22 9/57
19	10	9	23	0	0				0	25	5	8	0				2 7/14
20(a)	1	1	1	1		0			0	0	5	8	6	1	0	0	0
(b)									0	1	3	0	0				
21	132	0	235	68	0	91			109	5	3	6	0	0	0		210 1/26
22	1	1	1	1	1				1	130	11	276	1				3

EXERCISE 6

| | Blue | | | | Red | | | Value of |
1	2	3	4	1	2	3	4	Game
7	1	0	4					
17	7	0	12					
0	28	51	65	0	0	1	3	0
28	0	3	17					
0	0	1	1					

Index

A CATALOG OF SELECTED
DOVER BOOKS
IN ALL FIELDS OF INTEREST

A CATALOG OF SELECTED DOVER
BOOKS IN ALL FIELDS OF INTEREST

CONCERNING THE SPIRITUAL IN ART, Wassily Kandinsky. Pioneering work by father of abstract art. Thoughts on color theory, nature of art. Analysis of earlier masters. 12 illustrations. 80pp. of text. 5⅜ x 8½.   23411-8

ANIMALS: 1,419 Copyright-Free Illustrations of Mammals, Birds, Fish, Insects, etc., Jim Harter (ed.). Clear wood engravings present, in extremely lifelike poses, over 1,000 species of animals. One of the most extensive pictorial sourcebooks of its kind. Captions. Index. 284pp. 9 x 12.   23766-4

CELTIC ART: The Methods of Construction, George Bain. Simple geometric techniques for making Celtic interlacements, spirals, Kells-type initials, animals, humans, etc. Over 500 illustrations. 160pp. 9 x 12. (Available in U.S. only.)   22923-8

AN ATLAS OF ANATOMY FOR ARTISTS, Fritz Schider. Most thorough reference work on art anatomy in the world. Hundreds of illustrations, including selections from works by Vesalius, Leonardo, Goya, Ingres, Michelangelo, others. 593 illustrations. 192pp. 7⅛ x 10¼.   20241-0

CELTIC HAND STROKE-BY-STROKE (Irish Half-Uncial from "The Book of Kells"): An Arthur Baker Calligraphy Manual, Arthur Baker. Complete guide to creating each letter of the alphabet in distinctive Celtic manner. Covers hand position, strokes, pens, inks, paper, more. Illustrated. 48pp. 8¼ x 11.   24336-2

EASY ORIGAMI, John Montroll. Charming collection of 32 projects (hat, cup, pelican, piano, swan, many more) specially designed for the novice origami hobbyist. Clearly illustrated easy-to-follow instructions insure that even beginning papercrafters will achieve successful results. 48pp. 8¼ x 11.   27298-2

THE COMPLETE BOOK OF BIRDHOUSE CONSTRUCTION FOR WOOD-WORKERS, Scott D. Campbell. Detailed instructions, illustrations, tables. Also data on bird habitat and instinct patterns. Bibliography. 3 tables. 63 illustrations in 15 figures. 48pp. 5¼ x 8½.   24407-5

BLOOMINGDALE'S ILLUSTRATED 1886 CATALOG: Fashions, Dry Goods and Housewares, Bloomingdale Brothers. Famed merchants' extremely rare catalog depicting about 1,700 products: clothing, housewares, firearms, dry goods, jewelry, more. Invaluable for dating, identifying vintage items. Also, copyright-free graphics for artists, designers. Co-published with Henry Ford Museum & Greenfield Village. 160pp. 8¼ x 11.   25780-0

HISTORIC COSTUME IN PICTURES, Braun & Schneider. Over 1,450 costumed figures in clearly detailed engravings–from dawn of civilization to end of 19th century. Captions. Many folk costumes. 256pp. 8⅜ x 11¾.   23150-X

STICKLEY CRAFTSMAN FURNITURE CATALOGS, Gustav Stickley and L. & J. G. Stickley. Beautiful, functional furniture in two authentic catalogs from 1910. 594 illustrations, including 277 photos, show settles, rockers, armchairs, reclining chairs, bookcases, desks, tables. 183pp. 6½ x 9¼.   23838-5

AMERICAN LOCOMOTIVES IN HISTORIC PHOTOGRAPHS: 1858 to 1949, Ron Ziel (ed.). A rare collection of 126 meticulously detailed official photographs, called "builder portraits," of American locomotives that majestically chronicle the rise of steam locomotive power in America. Introduction. Detailed captions. xi+ 129pp. 9 x 12.   27393-8

AMERICA'S LIGHTHOUSES: An Illustrated History, Francis Ross Holland, Jr. Delightfully written, profusely illustrated fact-filled survey of over 200 American light-houses since 1716. History, anecdotes, technological advances, more. 240pp. 8 x 10¾.
25576-X

TOWARDS A NEW ARCHITECTURE, Le Corbusier. Pioneering manifesto by founder of "International School." Technical and aesthetic theories, views of industry, eco-nomics, relation of form to function, "mass-production split" and much more. Profusely illustrated. 320pp. 6⅛ x 9¼. (Available in U.S. only.)   25023-7

HOW THE OTHER HALF LIVES, Jacob Riis. Famous journalistic record, expos-ing poverty and degradation of New York slums around 1900, by major social reformer. 100 striking and influential photographs. 233pp. 10 x 7⅞.   22012-5

FRUIT KEY AND TWIG KEY TO TREES AND SHRUBS, William M. Harlow. One of the handiest and most widely used identification aids. Fruit key covers 120 deciduous and evergreen species; twig key 160 deciduous species. Easily used. Over 300 photographs. 126pp. 5⅜ x 8½.   20511-8

COMMON BIRD SONGS, Dr. Donald J. Borror. Songs of 60 most common U.S. birds: robins, sparrows, cardinals, bluejays, finches, more—arranged in order of increasing complexity. Up to 9 variations of songs of each species.
Cassette and manual 99911-4

ORCHIDS AS HOUSE PLANTS, Rebecca Tyson Northen. Grow cattleyas and many other kinds of orchids—in a window, in a case, or under artificial light. 63 illus-trations. 148pp. 5⅜ x 8½.   23261-1

MONSTER MAZES, Dave Phillips. Masterful mazes at four levels of difficulty. Avoid deadly perils and evil creatures to find magical treasures. Solutions for all 32 exciting illustrated puzzles. 48pp. 8¼ x 11.   26005-4

MOZART'S DON GIOVANNI (DOVER OPERA LIBRETTO SERIES), Wolfgang Amadeus Mozart. Introduced and translated by Ellen H. Bleiler. Standard Italian libretto, with complete English translation. Convenient and thoroughly portable—an ideal companion for reading along with a recording or the performance itself. Introduction. List of characters. Plot summary. 121pp. 5¼ x 8½.   24944-1

TECHNICAL MANUAL AND DICTIONARY OF CLASSICAL BALLET, Gail Grant. Defines, explains, comments on steps, movements, poses and concepts. 15-page pictorial section. Basic book for student, viewer. 127pp. 5⅜ x 8½.   21843-0

THE CLARINET AND CLARINET PLAYING, David Pino. Lively, comprehensive work features suggestions about technique, musicianship, and musical interpretation, as well as guidelines for teaching, making your own reeds, and preparing for public performance. Includes an intriguing look at clarinet history. "A godsend," *The Clarinet,* Journal of the International Clarinet Society. Appendixes. 7 illus. 320pp. 5⅜ x 8½. 40270-3

HOLLYWOOD GLAMOR PORTRAITS, John Kobal (ed.). 145 photos from 1926-49. Harlow, Gable, Bogart, Bacall; 94 stars in all. Full background on photographers, technical aspects. 160pp. 8⅜ x 11¼. 23352-9

THE ANNOTATED CASEY AT THE BAT: A Collection of Ballads about the Mighty Casey/Third, Revised Edition, Martin Gardner (ed.). Amusing sequels and parodies of one of America's best-loved poems: Casey's Revenge, Why Casey Whiffed, Casey's Sister at the Bat, others. 256pp. 5⅜ x 8½. 28598-7

THE RAVEN AND OTHER FAVORITE POEMS, Edgar Allan Poe. Over 40 of the author's most memorable poems: "The Bells," "Ulalume," "Israfel," "To Helen," "The Conqueror Worm," "Eldorado," "Annabel Lee," many more. Alphabetic lists of titles and first lines. 64pp. 5$\frac{3}{16}$ x 8¼. 26685-0

PERSONAL MEMOIRS OF U. S. GRANT, Ulysses Simpson Grant. Intelligent, deeply moving firsthand account of Civil War campaigns, considered by many the finest military memoirs ever written. Includes letters, historic photographs, maps and more. 528pp. 6⅛ x 9¼. 28587-1

ANCIENT EGYPTIAN MATERIALS AND INDUSTRIES, A. Lucas and J. Harris. Fascinating, comprehensive, thoroughly documented text describes this ancient civilization's vast resources and the processes that incorporated them in daily life, including the use of animal products, building materials, cosmetics, perfumes and incense, fibers, glazed ware, glass and its manufacture, materials used in the mummification process, and much more. 544pp. 6⅛ x 9¼. (Available in U.S. only.) 40446-3

RUSSIAN STORIES/RUSSKIE RASSKAZY: A Dual-Language Book, edited by Gleb Struve. Twelve tales by such masters as Chekhov, Tolstoy, Dostoevsky, Pushkin, others. Excellent word-for-word English translations on facing pages, plus teaching and study aids, Russian/English vocabulary, biographical/critical introductions, more. 416pp. 5⅜ x 8½. 26244-8

PHILADELPHIA THEN AND NOW: 60 Sites Photographed in the Past and Present, Kenneth Finkel and Susan Oyama. Rare photographs of City Hall, Logan Square, Independence Hall, Betsy Ross House, other landmarks juxtaposed with contemporary views. Captures changing face of historic city. Introduction. Captions. 128pp. 8¼ x 11. 25790-8

AIA ARCHITECTURAL GUIDE TO NASSAU AND SUFFOLK COUNTIES, LONG ISLAND, The American Institute of Architects, Long Island Chapter, and the Society for the Preservation of Long Island Antiquities. Comprehensive, well-researched and generously illustrated volume brings to life over three centuries of Long Island's great architectural heritage. More than 240 photographs with authoritative, extensively detailed captions. 176pp. 8¼ x 11. 26946-9

NORTH AMERICAN INDIAN LIFE: Customs and Traditions of 23 Tribes, Elsie Clews Parsons (ed.). 27 fictionalized essays by noted anthropologists examine religion, customs, government, additional facets of life among the Winnebago, Crow, Zuni, Eskimo, other tribes. 480pp. 6⅛ x 9¼. 27377-6

FRANK LLOYD WRIGHT'S DANA HOUSE, Donald Hoffmann. Pictorial essay of residential masterpiece with over 160 interior and exterior photos, plans, elevations, sketches and studies. 128pp. 9¼ x 10¾.   29120-0

THE MALE AND FEMALE FIGURE IN MOTION: 60 Classic Photographic Sequences, Eadweard Muybridge. 60 true-action photographs of men and women walking, running, climbing, bending, turning, etc., reproduced from rare 19th-century masterpiece. vi + 121pp. 9 x 12.   24745-7

1001 QUESTIONS ANSWERED ABOUT THE SEASHORE, N. J. Berrill and Jacquelyn Berrill. Queries answered about dolphins, sea snails, sponges, starfish, fishes, shore birds, many others. Covers appearance, breeding, growth, feeding, much more. 305pp. 5¼ x 8¼.   23366-9

ATTRACTING BIRDS TO YOUR YARD, William J. Weber. Easy-to-follow guide offers advice on how to attract the greatest diversity of birds: birdhouses, feeders, water and waterers, much more. 96pp. 5³⁄₁₆ x 8¼.   28927-3

MEDICINAL AND OTHER USES OF NORTH AMERICAN PLANTS: A Historical Survey with Special Reference to the Eastern Indian Tribes, Charlotte Erichsen-Brown. Chronological historical citations document 500 years of usage of plants, trees, shrubs native to eastern Canada, northeastern U.S. Also complete identifying information. 343 illustrations. 544pp. 6½ x 9¼.   25951-X

STORYBOOK MAZES, Dave Phillips. 23 stories and mazes on two-page spreads: Wizard of Oz, Treasure Island, Robin Hood, etc. Solutions. 64pp. 8¼ x 11.   23628-5

AMERICAN NEGRO SONGS: 230 Folk Songs and Spirituals, Religious and Secular, John W. Work. This authoritative study traces the African influences of songs sung and played by black Americans at work, in church, and as entertainment. The author discusses the lyric significance of such songs as "Swing Low, Sweet Chariot," "John Henry," and others and offers the words and music for 230 songs. Bibliography. Index of Song Titles. 272pp. 6½ x 9¼.   40271-1

MOVIE-STAR PORTRAITS OF THE FORTIES, John Kobal (ed.). 163 glamor, studio photos of 106 stars of the 1940s: Rita Hayworth, Ava Gardner, Marlon Brando, Clark Gable, many more. 176pp. 8⅜ x 11¼.   23546-7

BENCHLEY LOST AND FOUND, Robert Benchley. Finest humor from early 30s, about pet peeves, child psychologists, post office and others. Mostly unavailable elsewhere. 73 illustrations by Peter Arno and others. 183pp. 5⅜ x 8½.   22410-4

YEKL and THE IMPORTED BRIDEGROOM AND OTHER STORIES OF YIDDISH NEW YORK, Abraham Cahan. Film Hester Street based on *Yekl* (1896). Novel, other stories among first about Jewish immigrants on N.Y.'s East Side. 240pp. 5⅜ x 8½.   22427-9

SELECTED POEMS, Walt Whitman. Generous sampling from *Leaves of Grass*. Twenty-four poems include "I Hear America Singing," "Song of the Open Road," "I Sing the Body Electric," "When Lilacs Last in the Dooryard Bloom'd," "O Captain! My Captain!"—all reprinted from an authoritative edition. Lists of titles and first lines. 128pp. 5³⁄₁₆ x 8¼.   26878-0

THE BEST TALES OF HOFFMANN, E. T. A. Hoffmann. 10 of Hoffmann's most important stories: "Nutcracker and the King of Mice," "The Golden Flowerpot," etc. 458pp. 5⅜ x 8½. 21793-0

FROM FETISH TO GOD IN ANCIENT EGYPT, E. A. Wallis Budge. Rich detailed survey of Egyptian conception of "God" and gods, magic, cult of animals, Osiris, more. Also, superb English translations of hymns and legends. 240 illustrations. 545pp. 5⅜ x 8½. 25803-3

FRENCH STORIES/CONTES FRANÇAIS: A Dual-Language Book, Wallace Fowlie. Ten stories by French masters, Voltaire to Camus: "Micromegas" by Voltaire; "The Atheist's Mass" by Balzac; "Minuet" by de Maupassant; "The Guest" by Camus, six more. Excellent English translations on facing pages. Also French-English vocabulary list, exercises, more. 352pp. 5⅜ x 8½. 26443-2

CHICAGO AT THE TURN OF THE CENTURY IN PHOTOGRAPHS: 122 Historic Views from the Collections of the Chicago Historical Society, Larry A. Viskochil. Rare large-format prints offer detailed views of City Hall, State Street, the Loop, Hull House, Union Station, many other landmarks, circa 1904-1913. Introduction. Captions. Maps. 144pp. 9⅜ x 12¼. 24656-6

OLD BROOKLYN IN EARLY PHOTOGRAPHS, 1865-1929, William Lee Younger. Luna Park, Gravesend race track, construction of Grand Army Plaza, moving of Hotel Brighton, etc. 157 previously unpublished photographs. 165pp. 8⅞ x 11¾. 23587-4

THE MYTHS OF THE NORTH AMERICAN INDIANS, Lewis Spence. Rich anthology of the myths and legends of the Algonquins, Iroquois, Pawnees and Sioux, prefaced by an extensive historical and ethnological commentary. 36 illustrations. 480pp. 5⅜ x 8½. 25967-6

AN ENCYCLOPEDIA OF BATTLES: Accounts of Over 1,560 Battles from 1479 B.C. to the Present, David Eggenberger. Essential details of every major battle in recorded history from the first battle of Megiddo in 1479 B.C. to Grenada in 1984. List of Battle Maps. New Appendix covering the years 1967-1984. Index. 99 illustrations. 544pp. 6½ x 9¼. 24913-1

SAILING ALONE AROUND THE WORLD, Captain Joshua Slocum. First man to sail around the world, alone, in small boat. One of great feats of seamanship told in delightful manner. 67 illustrations. 294pp. 5⅜ x 8½. 20326-3

ANARCHISM AND OTHER ESSAYS, Emma Goldman. Powerful, penetrating, prophetic essays on direct action, role of minorities, prison reform, puritan hypocrisy, violence, etc. 271pp. 5⅜ x 8½. 22484-8

MYTHS OF THE HINDUS AND BUDDHISTS, Ananda K. Coomaraswamy and Sister Nivedita. Great stories of the epics; deeds of Krishna, Shiva, taken from puranas, Vedas, folk tales; etc. 32 illustrations. 400pp. 5⅜ x 8½. 21759-0

THE TRAUMA OF BIRTH, Otto Rank. Rank's controversial thesis that anxiety neurosis is caused by profound psychological trauma which occurs at birth. 256pp. 5⅜ x 8½. 27974-X

A THEOLOGICO-POLITICAL TREATISE, Benedict Spinoza. Also contains unfinished Political Treatise. Great classic on religious liberty, theory of government on common consent. R. Elwes translation. Total of 421pp. 5⅜ x 8½. 20249-6

MY BONDAGE AND MY FREEDOM, Frederick Douglass. Born a slave, Douglass became outspoken force in antislavery movement. The best of Douglass' autobiographies. Graphic description of slave life. 464pp. 5⅜ x 8½.   22457-0

FOLLOWING THE EQUATOR: A Journey Around the World, Mark Twain. Fascinating humorous account of 1897 voyage to Hawaii, Australia, India, New Zealand, etc. Ironic, bemused reports on peoples, customs, climate, flora and fauna, politics, much more. 197 illustrations. 720pp. 5⅜ x 8½.   26113-1

THE PEOPLE CALLED SHAKERS, Edward D. Andrews. Definitive study of Shakers: origins, beliefs, practices, dances, social organization, furniture and crafts, etc. 33 illustrations. 351pp. 5⅜ x 8½.   21081-2

THE MYTHS OF GREECE AND ROME, H. A. Guerber. A classic of mythology, generously illustrated, long prized for its simple, graphic, accurate retelling of the principal myths of Greece and Rome, and for its commentary on their origins and significance. With 64 illustrations by Michelangelo, Raphael, Titian, Rubens, Canova, Bernini and others. 480pp. 5⅜ x 8½.   27584-1

PSYCHOLOGY OF MUSIC, Carl E. Seashore. Classic work discusses music as a medium from psychological viewpoint. Clear treatment of physical acoustics, auditory apparatus, sound perception, development of musical skills, nature of musical feeling, host of other topics. 88 figures. 408pp. 5⅜ x 8½.   21851-1

THE PHILOSOPHY OF HISTORY, Georg W. Hegel. Great classic of Western thought develops concept that history is not chance but rational process, the evolution of freedom. 457pp. 5⅜ x 8½.   20112-0

THE BOOK OF TEA, Kakuzo Okakura. Minor classic of the Orient: entertaining, charming explanation, interpretation of traditional Japanese culture in terms of tea ceremony. 94pp. 5⅜ x 8½.   20070-1

LIFE IN ANCIENT EGYPT, Adolf Erman. Fullest, most thorough, detailed older account with much not in more recent books, domestic life, religion, magic, medicine, commerce, much more. Many illustrations reproduce tomb paintings, carvings, hieroglyphs, etc. 597pp. 5⅜ x 8½.   22632-8

SUNDIALS, Their Theory and Construction, Albert Waugh. Far and away the best, most thorough coverage of ideas, mathematics concerned, types, construction, adjusting anywhere. Simple, nontechnical treatment allows even children to build several of these dials. Over 100 illustrations. 230pp. 5⅜ x 8½.   22947-5

THEORETICAL HYDRODYNAMICS, L. M. Milne-Thomson. Classic exposition of the mathematical theory of fluid motion, applicable to both hydrodynamics and aerodynamics. Over 600 exercises. 768pp. 6⅛ x 9¼.   68970-0

SONGS OF EXPERIENCE: Facsimile Reproduction with 26 Plates in Full Color, William Blake. 26 full-color plates from a rare 1826 edition. Includes "The Tyger," "London," "Holy Thursday," and other poems. Printed text of poems. 48pp. 5¼ x 7.
        24636-1

OLD-TIME VIGNETTES IN FULL COLOR, Carol Belanger Grafton (ed.). Over 390 charming, often sentimental illustrations, selected from archives of Victorian graphics—pretty women posing, children playing, food, flowers, kittens and puppies, smiling cherubs, birds and butterflies, much more. All copyright-free. 48pp. 9¼ x 12¼.
        27269-9

PERSPECTIVE FOR ARTISTS, Rex Vicat Cole. Depth, perspective of sky and sea, shadows, much more, not usually covered. 391 diagrams, 81 reproductions of drawings and paintings. 279pp. 5⅜ x 8½. 22487-2

DRAWING THE LIVING FIGURE, Joseph Sheppard. Innovative approach to artistic anatomy focuses on specifics of surface anatomy, rather than muscles and bones. Over 170 drawings of live models in front, back and side views, and in widely varying poses. Accompanying diagrams. 177 illustrations. Introduction. Index. 144pp. 8⅜ x11¼. 26723-7

GOTHIC AND OLD ENGLISH ALPHABETS: 100 Complete Fonts, Dan X. Solo. Add power, elegance to posters, signs, other graphics with 100 stunning copyright-free alphabets: Blackstone, Dolbey, Germania, 97 more—including many lower-case, numerals, punctuation marks. 104pp. 8⅛ x 11. 24695-7

HOW TO DO BEADWORK, Mary White. Fundamental book on craft from simple projects to five-bead chains and woven works. 106 illustrations. 142pp. 5⅜ x 8. 20697-1

THE BOOK OF WOOD CARVING, Charles Marshall Sayers. Finest book for beginners discusses fundamentals and offers 34 designs. "Absolutely first rate . . . well thought out and well executed."–E. J. Tangerman. 118pp. 7¾ x 10⅝. 23654-4

ILLUSTRATED CATALOG OF CIVIL WAR MILITARY GOODS: Union Army Weapons, Insignia, Uniform Accessories, and Other Equipment, Schuyler, Hartley, and Graham. Rare, profusely illustrated 1846 catalog includes Union Army uniform and dress regulations, arms and ammunition, coats, insignia, flags, swords, rifles, etc. 226 illustrations. 160pp. 9 x 12. 24939-5

WOMEN'S FASHIONS OF THE EARLY 1900s: An Unabridged Republication of "New York Fashions, 1909," National Cloak & Suit Co. Rare catalog of mail-order fashions documents women's and children's clothing styles shortly after the turn of the century. Captions offer full descriptions, prices. Invaluable resource for fashion, costume historians. Approximately 725 illustrations. 128pp. 8⅜ x 11¼. 27276-1

THE 1912 AND 1915 GUSTAV STICKLEY FURNITURE CATALOGS, Gustav Stickley. With over 200 detailed illustrations and descriptions, these two catalogs are essential reading and reference materials and identification guides for Stickley furniture. Captions cite materials, dimensions and prices. 112pp. 6½ x 9¼. 26676-1

EARLY AMERICAN LOCOMOTIVES, John H. White, Jr. Finest locomotive engravings from early 19th century: historical (1804–74), main-line (after 1870), special, foreign, etc. 147 plates. 142pp. 11⅜ x 8¼. 22772-3

THE TALL SHIPS OF TODAY IN PHOTOGRAPHS, Frank O. Braynard. Lavishly illustrated tribute to nearly 100 majestic contemporary sailing vessels: Amerigo Vespucci, Clearwater, Constitution, Eagle, Mayflower, Sea Cloud, Victory, many more. Authoritative captions provide statistics, background on each ship. 190 black-and-white photographs and illustrations. Introduction. 128pp. 8⅞ x 11¾. 27163-3

LITTLE BOOK OF EARLY AMERICAN CRAFTS AND TRADES, Peter Stockham (ed.). 1807 children's book explains crafts and trades: baker, hatter, cooper, potter, and many others. 23 copperplate illustrations. 140pp. 4⅝ x 6.   23336-7

VICTORIAN FASHIONS AND COSTUMES FROM HARPER'S BAZAR, 1867–1898, Stella Blum (ed.). Day costumes, evening wear, sports clothes, shoes, hats, other accessories in over 1,000 detailed engravings. 320pp. 9⅜ x 12¼. 22990-4

GUSTAV STICKLEY, THE CRAFTSMAN, Mary Ann Smith. Superb study surveys broad scope of Stickley's achievement, especially in architecture. Design philosophy, rise and fall of the Craftsman empire, descriptions and floor plans for many Craftsman houses, more. 86 black-and-white halftones. 31 line illustrations. Introduction 208pp. 6½ x 9¼.   27210-9

THE LONG ISLAND RAIL ROAD IN EARLY PHOTOGRAPHS, Ron Ziel. Over 220 rare photos, informative text document origin (1844) and development of rail service on Long Island. Vintage views of early trains, locomotives, stations, passengers, crews, much more. Captions. 8⅞ x 11¾.   26301-0

VOYAGE OF THE LIBERDADE, Joshua Slocum. Great 19th-century mariner's thrilling, first-hand account of the wreck of his ship off South America, the 35-foot boat he built from the wreckage, and its remarkable voyage home. 128pp. 5⅜ x 8½.   40022-0

TEN BOOKS ON ARCHITECTURE, Vitruvius. The most important book ever written on architecture. Early Roman aesthetics, technology, classical orders, site selection, all other aspects. Morgan translation. 331pp. 5⅜ x 8½.   20645-9

THE HUMAN FIGURE IN MOTION, Eadweard Muybridge. More than 4,500 stopped-action photos, in action series, showing undraped men, women, children jumping, lying down, throwing, sitting, wrestling, carrying, etc. 390pp. 7⅞ x 10⅝.   20204-6 Clothbd.

TREES OF THE EASTERN AND CENTRAL UNITED STATES AND CANADA, William M. Harlow. Best one-volume guide to 140 trees. Full descriptions, woodlore, range, etc. Over 600 illustrations. Handy size. 288pp. 4½ x 6⅜.   20395-6

SONGS OF WESTERN BIRDS, Dr. Donald J. Borror. Complete song and call repertoire of 60 western species, including flycatchers, juncoes, cactus wrens, many more–includes fully illustrated booklet.   Cassette and manual 99913-0

GROWING AND USING HERBS AND SPICES, Milo Miloradovich. Versatile handbook provides all the information needed for cultivation and use of all the herbs and spices available in North America. 4 illustrations. Index. Glossary. 236pp. 5⅜ x 8½.   25058-X

BIG BOOK OF MAZES AND LABYRINTHS, Walter Shepherd. 50 mazes and labyrinths in all–classical, solid, ripple, and more–in one great volume. Perfect inexpensive puzzler for clever youngsters. Full solutions. 112pp. 8⅛ x 11.   22951-3

PIANO TUNING, J. Cree Fischer. Clearest, best book for beginner, amateur. Simple repairs, raising dropped notes, tuning by easy method of flattened fifths. No previous skills needed. 4 illustrations. 201pp. 5⅜ x 8½. 23267-0

HINTS TO SINGERS, Lillian Nordica. Selecting the right teacher, developing confidence, overcoming stage fright, and many other important skills receive thoughtful discussion in this indispensible guide, written by a world-famous diva of four decades' experience. 96pp. 5⅜ x 8½. 40094-8

THE COMPLETE NONSENSE OF EDWARD LEAR, Edward Lear. All nonsense limericks, zany alphabets, Owl and Pussycat, songs, nonsense botany, etc., illustrated by Lear. Total of 320pp. 5⅜ x 8½. (Available in U.S. only.) 20167-8

VICTORIAN PARLOUR POETRY: An Annotated Anthology, Michael R. Turner. 117 gems by Longfellow, Tennyson, Browning, many lesser-known poets. "The Village Blacksmith," "Curfew Must Not Ring Tonight," "Only a Baby Small," dozens more, often difficult to find elsewhere. Index of poets, titles, first lines. xxiii + 325pp. 5⅝ x 8¼. 27044-0

DUBLINERS, James Joyce. Fifteen stories offer vivid, tightly focused observations of the lives of Dublin's poorer classes. At least one, "The Dead," is considered a masterpiece. Reprinted complete and unabridged from standard edition. 160pp. 5$\frac{3}{16}$ x 8¼. 26870-5

GREAT WEIRD TALES: 14 Stories by Lovecraft, Blackwood, Machen and Others, S. T. Joshi (ed.). 14 spellbinding tales, including "The Sin Eater," by Fiona McLeod, "The Eye Above the Mantel," by Frank Belknap Long, as well as renowned works by R. H. Barlow, Lord Dunsany, Arthur Machen, W. C. Morrow and eight other masters of the genre. 256pp. 5⅜ x 8½. (Available in U.S. only.) 40436-6

THE BOOK OF THE SACRED MAGIC OF ABRAMELIN THE MAGE, translated by S. MacGregor Mathers. Medieval manuscript of ceremonial magic. Basic document in Aleister Crowley, Golden Dawn groups. 268pp. 5⅜ x 8½. 23211-5

NEW RUSSIAN-ENGLISH AND ENGLISH-RUSSIAN DICTIONARY, M. A. O'Brien. This is a remarkably handy Russian dictionary, containing a surprising amount of information, including over 70,000 entries. 366pp. 4½ x 6⅛. 20208-9

HISTORIC HOMES OF THE AMERICAN PRESIDENTS, Second, Revised Edition, Irvin Haas. A traveler's guide to American Presidential homes, most open to the public, depicting and describing homes occupied by every American President from George Washington to George Bush. With visiting hours, admission charges, travel routes. 175 photographs. Index. 160pp. 8¼ x 11. 26751-2

NEW YORK IN THE FORTIES, Andreas Feininger. 162 brilliant photographs by the well-known photographer, formerly with *Life* magazine. Commuters, shoppers, Times Square at night, much else from city at its peak. Captions by John von Hartz. 181pp. 9¼ x 10¾. 23585-8

INDIAN SIGN LANGUAGE, William Tomkins. Over 525 signs developed by Sioux and other tribes. Written instructions and diagrams. Also 290 pictographs. 111pp. 6⅛ x 9¼. 22029-X

ANATOMY: A Complete Guide for Artists, Joseph Sheppard. A master of figure drawing shows artists how to render human anatomy convincingly. Over 460 illustrations. 224pp. 8⅜ x 11¼. 27279-6

MEDIEVAL CALLIGRAPHY: Its History and Technique, Marc Drogin. Spirited history, comprehensive instruction manual covers 13 styles (ca. 4th century through 15th). Excellent photographs; directions for duplicating medieval techniques with modern tools. 224pp. 8⅜ x 11¼. 26142-5

DRIED FLOWERS: How to Prepare Them, Sarah Whitlock and Martha Rankin. Complete instructions on how to use silica gel, meal and borax, perlite aggregate, sand and borax, glycerine and water to create attractive permanent flower arrangements. 12 illustrations. 32pp. 5⅜ x 8½. 21802-3

EASY-TO-MAKE BIRD FEEDERS FOR WOODWORKERS, Scott D. Campbell. Detailed, simple-to-use guide for designing, constructing, caring for and using feeders. Text, illustrations for 12 classic and contemporary designs. 96pp. 5⅜ x 8½. 25847-5

SCOTTISH WONDER TALES FROM MYTH AND LEGEND, Donald A. Mackenzie. 16 lively tales tell of giants rumbling down mountainsides, of a magic wand that turns stone pillars into warriors, of gods and goddesses, evil hags, powerful forces and more. 240pp. 5⅜ x 8½. 29677-6

THE HISTORY OF UNDERCLOTHES, C. Willett Cunnington and Phyllis Cunnington. Fascinating, well-documented survey covering six centuries of English undergarments, enhanced with over 100 illustrations: 12th-century laced-up bodice, footed long drawers (1795), 19th-century bustles, 19th-century corsets for men, Victorian "bust improvers," much more. 272pp. 5⅝ x 8¼. 27124-2

ARTS AND CRAFTS FURNITURE: The Complete Brooks Catalog of 1912, Brooks Manufacturing Co. Photos and detailed descriptions of more than 150 now very collectible furniture designs from the Arts and Crafts movement depict davenports, settees, buffets, desks, tables, chairs, bedsteads, dressers and more, all built of solid, quarter-sawed oak. Invaluable for students and enthusiasts of antiques, Americana and the decorative arts. 80pp. 6½ x 9¼. 27471-3

WILBUR AND ORVILLE: A Biography of the Wright Brothers, Fred Howard. Definitive, crisply written study tells the full story of the brothers' lives and work. A vividly written biography, unparalleled in scope and color, that also captures the spirit of an extraordinary era. 560pp. 6⅛ x 9¼. 40297-5

THE ARTS OF THE SAILOR: Knotting, Splicing and Ropework, Hervey Garrett Smith. Indispensable shipboard reference covers tools, basic knots and useful hitches; handsewing and canvas work, more. Over 100 illustrations. Delightful reading for sea lovers. 256pp. 5⅜ x 8½. 26440-8

FRANK LLOYD WRIGHT'S FALLINGWATER: The House and Its History, Second, Revised Edition, Donald Hoffmann. A total revision–both in text and illustrations–of the standard document on Fallingwater, the boldest, most personal architectural statement of Wright's mature years, updated with valuable new material from the recently opened Frank Lloyd Wright Archives. "Fascinating"–*The New York Times*. 116 illustrations. 128pp. 9¼ x 10¾. 27430-6

PHOTOGRAPHIC SKETCHBOOK OF THE CIVIL WAR, Alexander Gardner. 100 photos taken on field during the Civil War. Famous shots of Manassas Harper's Ferry, Lincoln, Richmond, slave pens, etc. 244pp. 10⅝ x 8¼. 22731-6

FIVE ACRES AND INDEPENDENCE, Maurice G. Kains. Great back-to-the-land classic explains basics of self-sufficient farming. The one book to get. 95 illustrations. 397pp. 5⅜ x 8½. 20974-1

SONGS OF EASTERN BIRDS, Dr. Donald J. Borror. Songs and calls of 60 species most common to eastern U.S.: warblers, woodpeckers, flycatchers, thrushes, larks, many more in high-quality recording. Cassette and manual 99912-2

A MODERN HERBAL, Margaret Grieve. Much the fullest, most exact, most useful compilation of herbal material. Gigantic alphabetical encyclopedia, from aconite to zedoary, gives botanical information, medical properties, folklore, economic uses, much else. Indispensable to serious reader. 161 illustrations. 888pp. 6½ x 9¼. 2-vol. set. (Available in U.S. only.) Vol. I: 22798-7
Vol. II: 22799-5

HIDDEN TREASURE MAZE BOOK, Dave Phillips. Solve 34 challenging mazes accompanied by heroic tales of adventure. Evil dragons, people-eating plants, bloodthirsty giants, many more dangerous adversaries lurk at every twist and turn. 34 mazes, stories, solutions. 48pp. 8¼ x 11. 24566-7

LETTERS OF W. A. MOZART, Wolfgang A. Mozart. Remarkable letters show bawdy wit, humor, imagination, musical insights, contemporary musical world; includes some letters from Leopold Mozart. 276pp. 5⅜ x 8½. 22859-2

BASIC PRINCIPLES OF CLASSICAL BALLET, Agrippina Vaganova. Great Russian theoretician, teacher explains methods for teaching classical ballet. 118 illustrations. 175pp. 5⅜ x 8½. 22036-2

THE JUMPING FROG, Mark Twain. Revenge edition. The original story of The Celebrated Jumping Frog of Calaveras County, a hapless French translation, and Twain's hilarious "retranslation" from the French. 12 illustrations. 66pp. 5⅜ x 8½. 22686-7

BEST REMEMBERED POEMS, Martin Gardner (ed.). The 126 poems in this superb collection of 19th- and 20th-century British and American verse range from Shelley's "To a Skylark" to the impassioned "Renascence" of Edna St. Vincent Millay and to Edward Lear's whimsical "The Owl and the Pussycat." 224pp. 5⅜ x 8½. 27165-X

COMPLETE SONNETS, William Shakespeare. Over 150 exquisite poems deal with love, friendship, the tyranny of time, beauty's evanescence, death and other themes in language of remarkable power, precision and beauty. Glossary of archaic terms. 80pp. 5$\frac{3}{16}$ x 8¼. 26686-9

THE BATTLES THAT CHANGED HISTORY, Fletcher Pratt. Eminent historian profiles 16 crucial conflicts, ancient to modern, that changed the course of civilization. 352pp. 5⅜ x 8½. 41129-X

THE WIT AND HUMOR OF OSCAR WILDE, Alvin Redman (ed.). More than 1,000 ripostes, paradoxes, wisecracks: Work is the curse of the drinking classes; I can resist everything except temptation; etc. 258pp. 5⅜ x 8½. 20602-5

SHAKESPEARE LEXICON AND QUOTATION DICTIONARY, Alexander Schmidt. Full definitions, locations, shades of meaning in every word in plays and poems. More than 50,000 exact quotations. 1,485pp. 6½ x 9¼. 2-vol. set.
Vol. 1: 22726-X
Vol. 2: 22727-8

SELECTED POEMS, Emily Dickinson. Over 100 best-known, best-loved poems by one of America's foremost poets, reprinted from authoritative early editions. No comparable edition at this price. Index of first lines. 64pp. 5³⁄₁₆ x 8¼. 26466-1

THE INSIDIOUS DR. FU-MANCHU, Sax Rohmer. The first of the popular mystery series introduces a pair of English detectives to their archnemesis, the diabolical Dr. Fu-Manchu. Flavorful atmosphere, fast-paced action, and colorful characters enliven this classic of the genre. 208pp. 5³⁄₁₆ x 8¼. 29898-1

THE MALLEUS MALEFICARUM OF KRAMER AND SPRENGER, translated by Montague Summers. Full text of most important witchhunter's "bible," used by both Catholics and Protestants. 278pp. 6⅝ x 10. 22802-9

SPANISH STORIES/CUENTOS ESPAÑOLES: A Dual-Language Book, Angel Flores (ed.). Unique format offers 13 great stories in Spanish by Cervantes, Borges, others. Faithful English translations on facing pages. 352pp. 5⅜ x 8½. 25399-6

GARDEN CITY, LONG ISLAND, IN EARLY PHOTOGRAPHS, 1869–1919, Mildred H. Smith. Handsome treasury of 118 vintage pictures, accompanied by carefully researched captions, document the Garden City Hotel fire (1899), the Vanderbilt Cup Race (1908), the first airmail flight departing from the Nassau Boulevard Aerodrome (1911), and much more. 96pp. 8⅞ x 11¾. 40669-5

OLD QUEENS, N.Y., IN EARLY PHOTOGRAPHS, Vincent F. Seyfried and William Asadorian. Over 160 rare photographs of Maspeth, Jamaica, Jackson Heights, and other areas. Vintage views of DeWitt Clinton mansion, 1939 World's Fair and more. Captions. 192pp. 8⅞ x 11. 26358-4

CAPTURED BY THE INDIANS: 15 Firsthand Accounts, 1750-1870, Frederick Drimmer. Astounding true historical accounts of grisly torture, bloody conflicts, relentless pursuits, miraculous escapes and more, by people who lived to tell the tale. 384pp. 5⅜ x 8½. 24901-8

THE WORLD'S GREAT SPEECHES (Fourth Enlarged Edition), Lewis Copeland, Lawrence W. Lamm, and Stephen J. McKenna. Nearly 300 speeches provide public speakers with a wealth of updated quotes and inspiration—from Pericles' funeral oration and William Jennings Bryan's "Cross of Gold Speech" to Malcolm X's powerful words on the Black Revolution and Earl of Spenser's tribute to his sister, Diana, Princess of Wales. 944pp. 5⅜ x 8⅜. 40903-1

THE BOOK OF THE SWORD, Sir Richard F. Burton. Great Victorian scholar/adventurer's eloquent, erudite history of the "queen of weapons"—from prehistory to early Roman Empire. Evolution and development of early swords, variations (sabre, broadsword, cutlass, scimitar, etc.), much more. 336pp. 6⅛ x 9¼. 25434-8

AUTOBIOGRAPHY: The Story of My Experiments with Truth, Mohandas K. Gandhi. Boyhood, legal studies, purification, the growth of the Satyagraha (nonviolent protest) movement. Critical, inspiring work of the man responsible for the freedom of India. 480pp. 5⅜ x 8½. (Available in U.S. only.) 24593-4

CELTIC MYTHS AND LEGENDS, T. W. Rolleston. Masterful retelling of Irish and Welsh stories and tales. Cuchulain, King Arthur, Deirdre, the Grail, many more. First paperback edition. 58 full-page illustrations. 512pp. 5⅜ x 8½. 26507-2

THE PRINCIPLES OF PSYCHOLOGY, William James. Famous long course complete, unabridged. Stream of thought, time perception, memory, experimental methods; great work decades ahead of its time. 94 figures. 1,391pp. 5⅜ x 8½. 2-vol. set. Vol. I: 20381-6 Vol. II: 20382-4

THE WORLD AS WILL AND REPRESENTATION, Arthur Schopenhauer. Definitive English translation of Schopenhauer's life work, correcting more than 1,000 errors, omissions in earlier translations. Translated by E. F. J. Payne. Total of 1,269pp. 5⅜ x 8½. 2-vol. set. Vol. 1: 21761-2 Vol. 2: 21762-0

MAGIC AND MYSTERY IN TIBET, Madame Alexandra David-Neel. Experiences among lamas, magicians, sages, sorcerers, Bonpa wizards. A true psychic discovery. 32 illustrations. 321pp. 5⅜ x 8½. (Available in U.S. only.) 22682-4

THE EGYPTIAN BOOK OF THE DEAD, E. A. Wallis Budge. Complete reproduction of Ani's papyrus, finest ever found. Full hieroglyphic text, interlinear transliteration, word-for-word translation, smooth translation. 533pp. 6½ x 9¼. 21866-X

MATHEMATICS FOR THE NONMATHEMATICIAN, Morris Kline. Detailed, college-level treatment of mathematics in cultural and historical context, with numerous exercises. Recommended Reading Lists. Tables. Numerous figures. 641pp. 5⅜ x 8½. 24823-2

PROBABILISTIC METHODS IN THE THEORY OF STRUCTURES, Isaac Elishakoff. Well-written introduction covers the elements of the theory of probability from two or more random variables, the reliability of such multivariable structures, the theory of random function, Monte Carlo methods of treating problems incapable of exact solution, and more. Examples. 502pp. 5⅜ x 8½. 40691-1

THE RIME OF THE ANCIENT MARINER, Gustave Doré, S. T. Coleridge. Doré's finest work; 34 plates capture moods, subtleties of poem. Flawless full-size reproductions printed on facing pages with authoritative text of poem. "Beautiful. Simply beautiful."–*Publisher's Weekly.* 77pp. 9¼ x 12. 22305-1

NORTH AMERICAN INDIAN DESIGNS FOR ARTISTS AND CRAFTSPEOPLE, Eva Wilson. Over 360 authentic copyright-free designs adapted from Navajo blankets, Hopi pottery, Sioux buffalo hides, more. Geometrics, symbolic figures, plant and animal motifs, etc. 128pp. 8⅜ x 11. (Not for sale in the United Kingdom.) 25341-4

SCULPTURE: Principles and Practice, Louis Slobodkin. Step-by-step approach to clay, plaster, metals, stone; classical and modern. 253 drawings, photos. 255pp. 8⅛ x 11. 22960-2

THE INFLUENCE OF SEA POWER UPON HISTORY, 1660–1783, A. T. Mahan. Influential classic of naval history and tactics still used as text in war colleges. First paperback edition. 4 maps. 24 battle plans. 640pp. 5⅜ x 8½. 25509-3

THE STORY OF THE TITANIC AS TOLD BY ITS SURVIVORS, Jack Winocour (ed.). What it was really like. Panic, despair, shocking inefficiency, and a little heroism. More thrilling than any fictional account. 26 illustrations. 320pp. 5⅜ x 8½.
20610-6

FAIRY AND FOLK TALES OF THE IRISH PEASANTRY, William Butler Yeats (ed.). Treasury of 64 tales from the twilight world of Celtic myth and legend: "The Soul Cages," "The Kildare Pooka," "King O'Toole and his Goose," many more. Introduction and Notes by W. B. Yeats. 352pp. 5⅜ x 8½.
26941-8

BUDDHIST MAHAYANA TEXTS, E. B. Cowell and others (eds.). Superb, accurate translations of basic documents in Mahayana Buddhism, highly important in history of religions. The Buddha-karita of Asvaghosha, Larger Sukhavativyuha, more. 448pp. 5⅜ x 8½.
25552-2

ONE TWO THREE . . . INFINITY: Facts and Speculations of Science, George Gamow. Great physicist's fascinating, readable overview of contemporary science: number theory, relativity, fourth dimension, entropy, genes, atomic structure, much more. 128 illustrations. Index. 352pp. 5⅜ x 8½.
25664-2

EXPERIMENTATION AND MEASUREMENT, W. J. Youden. Introductory manual explains laws of measurement in simple terms and offers tips for achieving accuracy and minimizing errors. Mathematics of measurement, use of instruments, experimenting with machines. 1994 edition. Foreword. Preface. Introduction. Epilogue. Selected Readings. Glossary. Index. Tables and figures. 128pp. 5⅜ x 8½. 40451-X

DALÍ ON MODERN ART: The Cuckolds of Antiquated Modern Art, Salvador Dalí. Influential painter skewers modern art and its practitioners. Outrageous evaluations of Picasso, Cézanne, Turner, more. 15 renderings of paintings discussed. 44 calligraphic decorations by Dalí. 96pp. 5⅜ x 8½. (Available in U.S. only.)
29220-7

ANTIQUE PLAYING CARDS: A Pictorial History, Henry René D'Allemagne. Over 900 elaborate, decorative images from rare playing cards (14th–20th centuries): Bacchus, death, dancing dogs, hunting scenes, royal coats of arms, players cheating, much more. 96pp. 9¼ x 12¼.
29265-7

MAKING FURNITURE MASTERPIECES: 30 Projects with Measured Drawings, Franklin H. Gottshall. Step-by-step instructions, illustrations for constructing handsome, useful pieces, among them a Sheraton desk, Chippendale chair, Spanish desk, Queen Anne table and a William and Mary dressing mirror. 224pp. 8⅛ x 11¼.
29338-6

THE FOSSIL BOOK: A Record of Prehistoric Life, Patricia V. Rich et al. Profusely illustrated definitive guide covers everything from single-celled organisms and dinosaurs to birds and mammals and the interplay between climate and man. Over 1,500 illustrations. 760pp. 7½ x 10⅛.
29371-8

Paperbound unless otherwise indicated. Available at your book dealer, online at **www.doverpublications.com**, or by writing to Dept. GI, Dover Publications, Inc., 31 East 2nd Street, Mineola, NY 11501. For current price information or for free catalogues (please indicate field of interest), write to Dover Publications or log on to **www.doverpublications.com** and see every Dover book in print. Dover publishes more than 500 books each year on science, elementary and advanced mathematics, biology, music, art, literary history, social sciences, and other areas.